SpringerBriefs in Environmental Science

SpringerBriefs in Environmental Science present concise summaries of cutting-edge research and practical applications across a wide spectrum of environmental fields, with fast turnaround time to publication. Featuring compact volumes of 50 to 125 pages, the series covers a range of content from professional to academic. Monographs of new material are considered for the SpringerBriefs in Environmental Science series.

Typical topics might include: a timely report of state-of-the-art analytical techniques, a bridge between new research results, as published in journal articles and a contextual literature review, a snapshot of a hot or emerging topic, an in-depth case study or technical example, a presentation of core concepts that students must understand in order to make independent contributions, best practices or protocols to be followed, a series of short case studies/debates highlighting a specific angle.

SpringerBriefs in Environmental Science allow authors to present their ideas and readers to absorb them with minimal time investment. Both solicited and unsolicited manuscripts are considered for publication.

More information about this series at http://www.springer.com/series/8868

Alex Caveen • Nick Polunin • Tim Gray
Selina Marguerite Stead

The Controversy over Marine Protected Areas

Science Meets Policy

Springer

Alex Caveen
Seafish
Grimsby
United Kingdom

Nick Polunin
School of Marine Science and Technology
Newcastle University
Newcastle on Tyne
United Kingdom

Tim Gray
School of Geography, Politics and Sociology
Newcastle University
Newcastle on Tyne
United Kingdom

Selina Marguerite Stead
School of Marine Science & Technology
Newcastle University
Newcastle on Tyne
United Kingdom

ISSN 2191-5547 ISSN 2191-5555 (electronic)
ISBN 978-3-319-10956-5 ISBN 978-3-319-10957-2 (eBook)
DOI 10.1007/978-3-319-10957-2
Springer Cham Heidelberg New York Dordrecht London

Library of Congress Control Number: 2014948602

Printed on acid-free paper

Springer is part of Springer Science+Business Media (www.springer.com)

This book is dedicated to Ian and Pam Caveen

Acknowledgements

We would like to thank two UK research councils—the Natural Environment Research Council and the Economic and Social Research Council—for indirectly funding this project.

Executive Summary

This book is a contribution to a controversy which has pre-occupied marine governance across the world during the past 20 years, and shows little sign of resolution. This is the debate over whether marine reserves (MRs) are a better means of protecting commercial fish stocks and marine biodiversity than is conventional fisheries management (CFM), which includes quota restrictions, gear regulations, and minimum landing sizes, combined with multi-use marine protected areas (MUMPAs). The debate is between 'nature protectionists' (NPs) who argue for an extensive network of marine reserves (MRs) or no-take zones (NTZs) in which all fishing activity would be legally prohibited; and 'social conservationists' (SCs) who argue for CFM complemented by carefully selected spatial restrictions designed to protect spawning areas of target fish and biodiversity. This book has six objectives: (a) to explain the extraordinary speed with which the NP argument gathered pace to make MRs the most favoured global policy initiative in current marine management policy (Chap. 2); (b) to confirm the ascendancy of the MR model in the academic literature (Chap. 3); (c) to discuss whether scientific advocacy for MRs has exceeded the limits of scientific objectivity by introducing a pro-MR bias into the peer-review process (Chap. 4); (d) to examine the scientific credentials of the case for MRs (Chap. 5); (e) to test whether NP or SC discourses have prevailed in the recent designation of marine conservation zones (MCZs) in the UK (Chap. 6); and (f) to discuss the wider implications of the debate between NR and SC, including whether they can be reconciled in practice if not in principle (Chap. 7).

Contents

List of Abbreviations

ABNJ	areas beyond national jurisdiction
AC	advocacy coalition
AMSA	Australian Marine Sciences Association
BACI	before-after-control-impact
BDACI	before-during-after-control-impact
BIOT	British Indian Ocean Territory
BMR	Barbados Marine Reserve
CBD	Convention on Biological Diversity
CCAMLR	Convention for the Conservation of Antarctic Marine Living Resources
CEFAS	Centre for Environment, Fisheries and Aquaculture Science
CFC	chlorofluorocarbon
CFM	conventional fisheries management
CFMA	conventional fisheries management approach
CFP	Common Fisheries Policy
CPUE	catch-per-unit-effort
Defra	Department for Environment, Food and Rural Affairs
EBA	ecosystem-based approach
EBFM	ecosystem-based fisheries management
EBM	ecosystem-based management
ECHR	European Court of Human Rights
EDM	early day motion
EEZ	exclusive economic zone
EMS	European marine site
ENG	ecological network guidance
ENGO	environmental non-governmental organisation
EpC	epistemic community
EUNIS	European nature information system
FAO	Food and Agriculture Organization of the United Nations
GBRMPA	Great Barrier Reef Marine Protected Area
GCS	global citation score
GOC	Global Ocean Commission
HPMCZ	highly protected marine conservation zone

HSMPA	high seas marine protected area
ICCAT	International Commission for the Conservation of Atlantic Tunas
IFCA	Inshore Fisheries and Conservation Authority
IMO	International Maritime Organization
ISCZ	Irish sea conservation zone
ISI	Institute for Scientific Information
IUCN	International Union for Conservation of Nature
IUU	illegal, unreported, and unregulated fishing
IWC	International Whaling Commission
JNCC	Joint Nature Conservation Committee
LCS	local citation score
LSMR	large-scale marine reserve
Marinet	Marine Network of Friends of the Earth Local Groups
MARXAN	marine spatially explicit annealing
MCAA	Marine and Coastal Access Act
MCBI	Marine Conservation Biology Institute
MCZ	marine conservation zone
MCS	Marine Conservation Society
MEOW	marine ecoregions of the world
MMO	Marine Management Organisation
MNR	marine nature reserve
MPA	marine protected area
MPAC	Marine Protected Area Coalition
MR	marine reserve
MRAG	Marine Resources Assessment Group
MSFD	Marine Strategy Framework Directive
MSP	marine spatial planning
MSY	maximum sustainable yield
MUMPA	multi-use marine protected area
NAMMCO	North Atlantic Marine Mammal Commission
NAMPAN	North America Marine Protected Area Network
Natura	EU-wide network of nature protection areas established under the Habitats Directive
NCEAS	National Center for Ecological Analysis and Synthesis
NE	Natural England
NEAFC	North East Atlantic Fisheries Commission
NFFO	National Federation of Fishermen's Organisations
NGO	non-governmental organisation
NOAA	National Oceanographic and Atmospheric Administration
NTMR	no-take marine reserves
NP	nature protectionist/protectionism
NTA	no-take area
NTMPA	no-take MPA
NTZ	no-take zone
NWIFCA	North Western Inshore Fisheries Conservation Authority

OSPAR	Oslo Paris Convention for the Protection of the Marine Environment of the North-East Atlantic
PA	protected area
PISCO	Partnership for Interdisciplinary Studies of Coastal Oceans
PMSU	Prime Minister's Strategy Unit
PP	precautionary principle
RAMSAR	Ramsar Convention on Wetlands of International Importance
RCEP	Royal Commission on Environmental Pollution
REF	Research Excellence Framework
RFMO	Regional Fisheries Management Organisation
RSPB	Royal Society for the Protection of Birds
SAC	special areas of conservation
SAP	Science Advisory Panel
SC	social conservationist/conservationism
SPA	special protection area
SNCA	Statutory Nature Conservation Agency
SSSI	site of special scientific interest
SST	sea surface temperature
TAN	transnational advocacy network
TBMPA	transboundary marine protected area
TEK	traditional ecological knowledge
UNCLOS	United Nations Convention on the Law of the Sea
UNEP	United Nations Environment Programme
WDCS	Whale and Dolphin Conservation Society
WoS	Web of Science
WSSD	World Summit on Sustainable Development
WWF	World Wide Fund for Nature

List of Figures

List of Tables

Chapter 1
Introduction

The most important controversy in fisheries management in recent years has been the debate over marine protected areas (MPAs). The epicentre of this controversy is the issue of whether large networks of no-take MPAs (NTMPAs), more usually known as marine reserves (MRs), are necessary to protect fish stocks. On the one hand, advocates of MRs argue that without MRs the worldwide decline in fish stocks will continue to the point of threatening more stocks with extinction. On the other hand, critics of MRs argue that conventional fisheries management (CFM) which includes restrictions on quota, fishing gear, effort controls, and selective spatial restrictions, if properly enforced, is perfectly adequate to protect fish stocks. There is another element in this controversy—which concerns marine biodiversity. Advocates of MRs argue that the only way to protect non-target marine species and their habitats (biodiversity) is to establish large networks of MRs; whereas critics of MRs argue that biodiversity can be adequately protected by CFM together with some selected MRs.

Miller et al. (2011, p. 948, 952) have depicted this conflict as between "nature protectionists" (NPs) and "social conservationists" (SCs):

> In one corner, are what might be called "nature protectionists", or conservation scientists and scholarly allies in fields such as environmental philosophy who defend protected areas (PAs) and conservation policies that strictly limit human presence and who advance biodiversity protection as the primary goal of international conservation efforts…In the other, are "social conservationists" who advocate various forms of sustainable use and privilege conservation-oriented development and welfare-oriented goals such as poverty alleviation and social justice…NPs generally conceive of humans as a threat to strict biodiversity conservation…while SCs believe that humans…can be allies in the conservation effort if incorporated effectively in park planning and management

Although the Miller et al. (2011) analysis was conducted in relation to terrestrial protected areas (the parks versus people debate), we will use their terminology to exemplify the MR controversy as a debate between NPs and SCs. Jones (2002) characterised this division as top-down versus bottom-up; science-based versus science-guided; principled versus pragmatic, and emphasised the ethical divide between NPs and SCs: "NTMPA proponents being more influenced by preservationist

A. Caveen et al., *The Controversy over Marine Protected Areas,*
SpringerBriefs in Environmental Science, DOI 10.1007/978-3-319-10957-2_1

and ecocentric perspectives, and CFMA proponents being more influenced by the utilitarian resource conservation perspective" (Jones 2007, p. 38) (see also Hilborn 2007c; Agardy et al. 2003).

The controversy between NPs and SCs has, therefore, two dimensions: an empirical dimension and a normative dimension. The empirical dimension is a factual dispute over whether extensive networks of MRs are necessary to protect fish stocks and biodiversity. Here both NPs and SCs agree on the objective (to protect fish stocks and biodiversity), but they disagree about the means to achieve that objective: NPs hold that extensive MR networks are necessary, whereas SCs hold that they are not necessary. This empirical dispute is potentially resolvable if sufficient data become available, or if a compromise can be reached between NPs and SCs to agree on some MRs. The normative dimension is an ethical dispute over whether marine resources should be preserved or utilised. Here NPs and SCs do not agree on the objective (whether to preserve or use marine resources), and so this normative dispute may never be resolved, unless one side persuades the other to change its value system. In this book, we will see how the controversy between NPs and SCs shifts confusingly between its empirical dimension and its normative dimension, making it difficult to predict whether a resolution between them will ever emerge.

1.1 The NP Argument

1.1.1 Empirical Dimension

The foundation of the nature protectionist argument is a pessimistic assessment of the state of the world's fish stocks. A growing body of evidence has documented the declining abundance and diversity of marine resources (Worm et al. 2009) and the negative effects of fishing on marine ecosystems (Agardy 2000). Many marine species have become extinct (Jackson et al. 2001) or are in the process of becoming extinct (Roberts and Hawkins 1999), and there have been significant declines in large predatory fish (Pauly et al. 2002; Myers and Worm 2003). According to the latest Food and Agriculture Organisation (FAO) (2012) report, as of 2009, globally 29.9% of fish stocks are overexploited[1], 57.4% are fully exploited, and 12.7% are not fully exploited. Many scientists argue that we are facing a fisheries crisis (Roberts 1997), with massive implications for long-term food security (Pauly et al. 2002; Smith et al. 2010; Godfray et al. 2010). Moreover, overfishing has destroyed habitats (Dayton et al. 1995) and altered marine ecosystems either directly (Watling and Norse 1998) or indirectly (Pinnegar et al. 2000; Baum and Worm 2009). Koldewey et al. (2010, p. 1910) claimed that "Fisheries are the largest anthropogenic threat to pelagic ecosystems, therefore preventing fishing will potentially have the greatest beneficial effect for the ecosystem".

[1] According to some, this is likely to be an underestimate. Pauly and Froese (2012) suggested that 37% of fish stocks yield less than 10% of their historic maximum catches.

So the empirical dimension of the NP argument is that conventional fisheries management (CFM) has failed and that radical new solutions such as MRs are needed to reverse the decline in abundance and biodiversity of marine resources (Halpern 2003; Roberts et al. 2005; Beare et al. 2014). As Kaiser (2005, p. 1194) put it, "MPAs have been heralded as the saviour of global fisheries by some conservationists, fishers and managers and are seen as the solution to the perceived failures of current management methods" (see also Cvitanovic et al. 2013). Many studies purport to show that MRs increase commercial and other fish stocks (Mosquera et al. 2000), and benefit non-target species and habitats (Micheli et al. 2004; Lester et al. 2009; Russ and Alcala 2011). Halpern (2003, p. S117, S129) claimed that a review of the literature showed that "nearly any marine habitat can benefit from the implementation of a reserve…marine reserves, regardless of their size, and with few exceptions, lead to increases in density, biomass, individual size, and diversity in all functional groups" (see also Tissot et al. 2013; Lubchenco et al. 2003; Metcalfe 2013). Lubchenco et al. (2007, p. 4) asserted that "Scientists have studied more than 124 marine reserves around the world and monitored biological changes inside the reserves…nearly all the effects were positive". Laffoley (2012) claimed that "there is no better tool for recovering marine biodiversity than marine reserves. No matter what anyone tells you anywhere the reality is that when you put in place a marine reserve there are a few examples where there have been negative effects, but generally on average it is around 440% increase in biomass and up to several thousand increases in some cases of biodiversity within an area". According to NPs, MRs protect unique underwater features, biodiversity hotspots, and threatened or rare species (Kelleher et al. 1995; Farrow 1996; Grafton et al. 2011). MRs established to conserve nature may also have wider societal value such as marine education and scientific research (Leisher et al. 2012), e.g. as control areas (Laffoley 2012). Moreover, NPs argue that while partially protected areas such as multi-use MPAs (MUMPAs) may be better than no protection at all, they are not as effective as MRs (NTMPAs) (Jones 2014; NCEAS 2001; Grorud-Colvert et al. 2011; Shears et al. 2006).

NPs have called for policy makers to establish very large MRs—a call that has been characterised as a return to the fortress conservation paradigm (De Santo et al. 2011). But it is actually a link to the ecosystem-based approach (EBA) to fisheries management. EBA is contrasted with single-species management because it entails that fisheries management is not about maximising the sustainable yield (MSY) of targeted fish stocks, but about protecting the health of a whole ecosystem, which includes fish species, benthic organisms, sea mammals, marine flora, and their natural habitats (Gray and Hatchard 2008; Fogarty and Murawski 1998). The NP justification of MRs is often linked to the EBA (or ecosystem-based management (EBM) (Fraschetti et al. 2011; Lubchenco et al. 2003; Mangi et al. 2011; Jones 2007) because MRs are holistic entities designed for the protection of entire ecosystems. Indeed, according to Angulo-Valdes and Hatcher (2010), MRs are the main tools for the implementation of EBA, and the Rio Earth Summit "actively promoted the importance of protected areas as a tool to implement ecosystem-based management" (Spalding et al. 2013, p. 216). The larger the MR, the bigger is the ecosystem that it protects, restores, recovers or rebuilds, and the more extensive the network

Table 1.1 Some high profile studies that made recommendations for MPAs to achieve EBM. Citations data correct on the 20th April 2012, taken from the Institute for Scientific Information (ISI)'s Web of science

Study	Citations	Recommendations for MPAs
Jackson et al. (2001)	1706	Large-scale, adaptive experiments for ecosystem restoration, exploitation, and management
Pauly et al. (1998)	1332	In the next decades fisheries management will have to emphasise the rebuilding of fish populations embedded in functional food webs, within large "no-take" marine protected areas
Pauly et al. (2002)	736	Zoning the oceans into unfished marine reserves and areas with limited levels of fishing effort would allow sustainable fisheries, based on resources embedded in functional, diverse ecosystems
Worm et al. (2006)	673	By restoring marine biodiversity through sustainable fisheries management, pollution control, maintenance of essential habitats, and the creation of marine reserves, we can invest in the productivity and reliability of the goods and services that the ocean provides to humanity
Conover and Munch (2002)	314	The establishment of no-take reserves or marine protected areas may, if properly designed, provide for the maintenance of natural genetic variation by allowing a proportion of the stock to develop an unconstrained range of size and growth rates

of MRs, the greater the ecological returns (Grorud-Colvert et al. 2011). Many of the most highly cited studies published in the marine science journals suggest that there ought to be networks of MRs explicitly or implicitly established to restore ecosystems (Ballantine 2014; Jones 2014; see Table 1.1). Networks of MPAs would produce greater ecological and socio-economic returns than would single MPAs (Grorud-Colvert et al. 2011; Gaines et al. 2010b), and Olsen et al. (2013) reported that some countries are introducing networks of MPAs as part of larger frameworks of EBA. Many NP scientists called for large areas of the world's oceans to be designated as MRs. For example, Roberts said "I would like to see 30% of the ocean no-take. I think that is justified by the scientific evidence" (Roberts and Hilborn 2013, p. 2).

A report by the United Nations Environment Programme (UNEP) suggested that many successful examples of EBM have originated from MPAs "because the discrete nature of protected areas allows *experimentation with EBM approaches and integration*—and [MPAs] often represent where the first steps along the EBM journey are taken" (UNEP 2011, p. 53; italics in original). This links to the characteristic of NP that argues for MRs as control sites for scientific research. As Ballantine (2014, p. 6, 8) put it: "In scientific terms a marine reserve is a "control", the unmanipulated part of an experimental design. This means that quite different scientific rules apply…We do not expect "responses" from controls in an experiment…they are essential to marine science, in the same sense that clean apparatus and pure reagents are essential to chemistry" (see also Olsen et al. 2013). NPs readily invoke the precautionary principle (PP) where data are thin (Mosquera et al. 2000).

The empirical NP case for MRs has been made not only on ecological grounds (protecting biodiversity including commercial fish stocks) but also on economic grounds. Balmford et al. (2004, p. 9697) calculated that a global network of MRs would cost annually US$5–19 billion to run, but yield an annual global fish catch of US$70–80 billion and annually deliver "largely unseen marine ecosystem services" of US$4.5–6.7 trillion. Ballantine (2014) pointed out that MRs have become major tourism attractions, generating substantial revenues for local economies. In a conference talk, Sala (2010) strikingly stated the NP's economic case for MRs: "What we have now—a world without marine reserves—is like a debit account where we withdraw all the time and we never make any deposit. Reserves are like savings accounts."

1.1.2 Normative Dimension

The normative dimension of the NP argument rests on the ethical grounds of preserving unique living creatures from extinction, which entails returning to a natural state before human exploitation. This dimension lies behind the writings of the well-known MR advocate, Bill Ballantine, of New Zealand. For instance, Ballantine (2014, p. 3–4, 6, 9) wrote that

> The essential regulations for marine reserves are those needed to maintain the full expression of the intrinsic processes in the sea and hence allow the free development of natural biodiversity at all levels…Marine reserves are kept free of all direct extractive or other disturbances on principle…Most no-take reserves have an underlying purpose to keep the environment in a more natural state than adjacent areas…What we need is not more data, better calculations and more micro-management, but…a significant proportion of no-take marine reserves that would allow the resumption of more natural dynamics (which did sustain all species before we came along).

1.2 The Social Conservationist Argument

1.2.1 Empirical Dimension

By contrast, under a SC framework, the NP approach is criticised for being too pessimistic about declining fish stocks and too optimistic about the value of MRs in reversing that decline. According to SC voices, the proposition that global fish stocks were crashing and that there was a danger of imminent species extinction, was misleading and exaggerated (Hilborn 2007b), not helped by sensationalistic science (Worm et al. 2006), sloppy journalism (Leake 2012), and endorsement by high profile conservationists[2]. On declining fish stocks, as O'Sullivan and Emmerson (2011, p. 116) put it, "there is a real danger of overestimating threats and

[2] For example at the end of the marine conservation zone (MCZ) planning process, the famous broadcaster Sir David Attenborough, vice president of the Wildlife Trusts said "I urge the government to designate the full list of 127 sites now, for day by day the wildlife in these sites is being

sensationalizing the process of extinction". Many scientists are wary about making gross generalisations about fish-stock declines and of laying the blame for their alleged decline entirely on conventional fisheries science and management tools[3]. They acknowledge that there are many failed fisheries, but they also point to some successes (Hilborn 2007a, b). Hilborn and Ovando (2014, p. 1040, 1045) claimed that the real failure is not fisheries management but the lack of fisheries management: "Stocks that are managed are improving, while stocks that are not managed are not…The evidence is strong that where fisheries management has been applied, it has worked to both reduce fishing pressure and to rebuild stocks…it is not the failure but the lack of management that drives fishery depletion". And SCs argue that "the apocalypse that many marine ecologists are warning of…has already been forestalled by improvements in CFMAs" (Jones 2014, p. 47).

On being too gung ho about MRs, Polunin et al. (2009, p. 6) stated that "the benefits of MPAs have too often been assumed despite being based on uncertain understanding of fish behaviour and ecology, and of fisheries themselves". Agardy et al. (2003, p. 354, 359) warned of the danger of exaggerating the benefits of MRs, referring to "the tendency to decree as many MPAs as possible, an eagerness to do so without a clear understanding of many of the complexities or balanced framework required, and a zealous "one-size-fits-all" approach…[and an] assertive promotion of no-take MPAs as the best and only effective type of MPA". Dunne et al. (2014, p. 24) pointed out that "direct evidence to support the creation of MPAs for fishing benefits is generally very limited and many arguments are in fact supported by nothing more than normative assumptions" (see also Stewart et al. 2008). Even in the case of coral reefs where the evidence for the role of MRs in mitigating impacts and aiding recovery within their borders is greatest, Sale et al. (2005, p. 74) held that "there are significant gaps in scientific knowledge that must be filled if no-take reserves are to be used effectively as fishery management tools. Unfortunately, these gaps are being glossed over by some uncritical advocacy".

Some SCs ask whether an MR network is necessary at all, or whether other fisheries management measures would suffice. Sweeting and Polunin (2005, p. 55) claimed that "habitat protection can be achieved by exclusion of benthic gears, avoidance of conflict can be achieved by spatial segregation and sustainable exploitation of fish stock can occur within trawl exclusions, a fact that is recognised by most MPA practitioners". Pajaro et al. (2010, p. 960) pointed out that "MPA establishment and management is a massive experiment in human environment relations". Is the experiment worth it? It has huge opportunity costs: As Fraschetti et al. (2011, p. 13) noted, "the tendency to use MPAs as the preferred management tool may preclude consideration of other management options". Hilborn (2013, p. 111) held that in "regulated fisheries that typify developed countries it is well demonstrated

destroyed and damaged. Time is running out for us to save our fragile seas". http://www.independent.co.uk/environment/nature/attenborough-issues-plea-to-save-our-seas-7834223.html.

[3] Indeed, the findings of many of these high impact studies have been alleged to be grossly misleading: "Closer inspection of this litany of papers shows them to be either outright wrong or serious distortions of reality" (Hilborn 2007d, p. 297).

that overfishing can be controlled and good biological outcomes achieved without areas permanently closed to fishing". Indeed, Suuronen et al. (2010, p. 243) stated that many modelling studies showed that "yields are maximised with effort regulation rather than by the introduction of closures" (see also Greenstreet et al. 2009). Likewise, Branch wrote that "numerous papers show that MPAs won't improve fisheries yield unless the populations are already overfished. In such cases yield would also improve if overfishing was reversed and rebuilding allowed ('traditional management')" (Roberts and Hilborn 2013, p. 7). Similarly, Kaiser (2005, p. 1198) wrote that "the proper implementation of fishing effort reduction still has the potential to out-perform the use of MPAs in terms of increasing spawning stock biomass" (see also Jones 2007; Kaiser 2004; Metcalfe 2013).

SCs also criticise NPs for ignoring the socio-economic harm caused by MRs. For example, Christie (2004) pointed out that MPAs which are regarded as biological successes may be socio-economic failures by increasing social conflict between marine users, and causing economic dislocation in disadvantaged artisanal fishing communities. Hilborn added the suggestion that "the vast funds and energy going into MPA establishment in developed countries would better be applied to improving fisheries management in places that do not have good management systems" (Roberts and Hilborn 2013, p. 3).

Another SC criticism of MRs is that they are invariably poorly enforced—"paper parks" (Jones 2014; Marinesque et al. 2012; Pomeroy et al. 2005; Agardy et al. 2011). McClanahan et al. (2006, p. 1408) claimed that where there was ineffective management, MUMPAs fared better than MRs: "In cases where the resources for enforcement are lacking, management regimes that are designed to meet community goals can achieve greater compliance and subsequent conservation success than regimes designed primarily for biodiversity conservation".

SCs also have doubts about MR networks. Roff (2014) claimed that there is no evidence that any MR networks have been implemented, and even when implemented, it will be extremely difficult to determine their effects, because there will be so many confounding factors. Moreover, SCs reject target percentages such as 30% of the oceans as MRs are scientifically unjustified (Planes 2011). Hilborn drew attention to the wider issue of whether large MR networks have a net global environmental benefit, in particular because of their displacement of food production from the seas to the land:

> Imagine we closed all the oceans to fishing. There is no doubt that fish abundance would rise and most measures of biodiversity status would improve in the oceans. But what else would happen?… If 80 million tons of lost fish production was made up by chopping down rain forest to grow cattle, I think the global environment would be worse off. Capture fisheries produce food at lower environmental cost than livestock and most forms of aquaculture. No water, pesticides, fertilizer or antibiotics are used, and greenhouse gas omissions are lower. These tradeoffs must be considered when large areas of the ocean are closed. (Roberts and Hilborn 2013, p. 1)

SCs also criticise the use of MRs for ecosystem-based fisheries management (EBFM). For example, Greenstreet et al. (2009) claimed that the displacement of effort from within an MR could lead to a net loss for the wider ecosystem, as in

Table 1.2 Five potential shortfalls of MPAs in relation to EBFM. (Agardy et al. 2011)

Shortcoming	References
Are ecologically insufficient by virtue of their small size or poor design	Bloomfield et al. (2012)
Are inappropriately planned or managed	Gerhardinger et al. (2011)
Fail due to the degradation of the unprotected surrounding ecosystems	Hale (2014)
Do more harm than good due to displacement and unintended consequences of management	Greenstreet et al. (2009); Abbott and Haynie (2012)
Create an illusion of protection when in fact no protection is occurring	Kareiva (2006)

the case of the North Sea "cod box", which was designed to reduce juvenile-cod mortality, but displaced effort to virgin areas where benthic habitat damage was inflicted. Moreover, the concept of restoring ecosystem health is problematic. For example, Lackey (2001, p. 440) argued that it is not a scientific term but a "highly charged political term", surrounded by controversy. It has no inherent (objective) meaning, but is dependent on subjective value judgements. One such value judgement is that "there is a 'natural' ecosystem state (i.e. balance of nature)", any deviation from which is deemed to signify ecosystem ill health (Lackey 2001, p. 442). There is an implicit assumption here that "an 'undisturbed' or 'natural' ecosystem is superior, thus preferred, to an 'altered' one…but there is nothing *scientific* that compels *any* specific ecological state to be considered preferred or better (more healthy)…expressions of ecosystem health reflect values and preferences" (Lackey 2001, p. 444–445; italics in original). In other words, there is no such "natural ecosystem state" out there, only an idealised notion in the minds of scientists who adumbrate it—a social construction.

Moreover, as Hilborn (2011, p. 236) points out, "if governments and fisheries agencies have been unsuccessful at implementing single-species management, should we expect them to successfully implement a necessarily more complex EBFM?" Furthermore, in most cases costs may be prohibitively too high to achieve full EBFM (Hilborn 2011). The key question not asked by the UK government during the planning of marine conservation zones (MCZs) in England was: Do we have the information, money, and time to attempt systematic conservation planning along the lines of EBFM, or should we be less ambitious and prioritise sites for protection that are known to be vulnerable to fishing by working more closely with the fishing industry (see Chap. 6)? In any case, MRs in themselves will never be able to deliver full EBFM, because, as Halpern et al. (2010, p. 18313) puts it, "even in the best case scenario MPAs can address only a subset of EBM goals" (see Table 1.2).

1.2.2 Normative Dimension

Behind these SC challenges to the empirical claims made by NPs lies a normative challenge. SCs argue for the ethical value of sustainable development against the

NP's intrinsic value of nature. For SCs, natural resources are valuable for their instrumental worth to humans, not their intrinsic value (however that is defined). As Pim Visser of Dutch producers' organisation VisNed put it, "These depend on your view of nature. Do we want to return to a wilderness in pristine condition, or do we want a nature that is productive as we care for it?" (*Fishing News* 29.4.11, p. 12). SCs are very concerned to protect biodiversity, not because every living creature has a right to life, but because certain creatures play important roles in conserving the ecosystem which provides so many important services to humanity. SCs objected to the NPs' idea that large areas of the sea should be set aside for scientific experiment: "scientist keen to preserve a 'pristine' natural laboratory" (Dunne et al. 2014, p. 24). Hilborn (Roberts and Hilborn 2013, p. 2) claimed that the issue of MRs was essentially ethical, not scientific: "I don't see this question as a scientific one—it is question of personal choice and my opinion has no more value than anyone else's. I believe it is clear that there is a trade-off between how much of the ocean we close, and food production".

SCs' ethical doubts about MRs also raised issues of equity or social justice, such as depriving fishers of their livelihoods without sufficient justification. For example, MRs may be imposed without proper consultation, with inadequate scientific data, in disproportionately great numbers, in unnecessarily rich fishing grounds, or in overseas territories such as Chagos which raised the charge of a "return to a neo-colonialist 'fortress conservation' approach…that neglects the social justice implications of such very large MPA designations" (Jones 2014, pp. 31–32). On the last point, De Santo et al. (2011) explained that in 2010 the British government's declaration of the world's largest MR of 210,000 miles^2 surrounding the Chagos Islands in the British Indian Ocean Territory (BIOT) ignored the fact that a legal case was pending in the European Court of Human Rights (ECHR) to judge whether the Chagossian islanders had a right to return to the islands after their forcible removal 40 years ago to make way for a military airbase.

This is not to say that SCs reject MRs out of hand. On the contrary, SCs see an important role for MRs as a conservation measure to protect essential fish habitats (Botsford et al. 2003), and to reduce fishing mortality on aggregations of spawning and feeding adult (Chiappone and Sealey 2000) and undersized juvenile fish (Schopka 2007). In certain circumstances MRs may also be used to enhance fisheries yields (Russ et al. 2004), though empirical evidence for this effect is sparse. However, SCs are much more in favour of MUMPAs than MRs, where the evidence shows that certain fishing activities within MUMPAs do not cause irreparable damage (Agardy et al. 2003; De Santo 2013; Hilborn et al. 2004; Guidetti and Claudet 2010). According to Claudet et al. (2011, p. 40), "Partial protection provided by multiple use MPAs…may, in some cases, confer the same benefits as full protection provided by no-take zones".

Under a SC framework, the main objective of MUMPAs is to protect the marine environment in order to provide ecosystem services. Ecosystem services include fish stock enhancement, poverty reduction, coastal protection, recreation, tourism, and carbon sequestration (Spalding et al. 2013). The term "ecosystem services" was incorporated at the Convention on Biological Diversity (CBD) meeting in Nagoya in 2010 into the Aichi Target 11 (Spalding et al. 2013). This was a critical shift in the CBD's focus from purely biological criteria to include socio-economic criteria: Instead of assuming that socio-economic benefits would automatically follow from biodiversity protection, the requirement was that MPAs must be explicitly designed to directly provide socio-economic benefits. The implications of this shift for MPA site selection are twofold: (1) increasing access for users to MPAs and (2) siting more MPAs inshore where ecosystem services are more accessible. Rees et al. (2014) argued that stressing the contribution that MPA networks could make to delivering ecosystem services would help to speed up the establishment of MPA networks across the world, though Potts et al. (2014) said that so far few MPA designations have explicitly taken ecosystem services into account.

One of the main ecosystem services is fisheries, and MRs are often proposed as a risk management strategy to serve as a buffer against uncertainty in fish stocks (Lauck et al. 1998; Clark 1996) for two reasons: (1) conventional management through catch or effort controls may fail due to stock assessment errors and an inadequate institutional framework (Finlayson 1994; Daw and Gray 2005) and (2) the functional roles a habitat and associated species assemblage have in contributing to valued ecosystem productivity are often not known (Frid and Paramor 2006). Given these uncertainties, some scientists argue that it is wise to designate MRs in order to protect part of a fish stock from exploitation (Lauck et al. 1998), as well as habitats and species that are sensitive to fishing (Watling and Norse 1998). However, the capacity of MRs to mitigate the unpredictable effects of management mistakes or the unforeseen natural loss of essential habitats/species is often undermined by the lack of relevant biological knowledge to design MRs to meet these objectives (Hilborn et al. 2004; Osenberg et al. 2011). According to Tissot et al. (2013, p. 115), "Few MPA studies have shown direct connections between MPA effectiveness and fishery benefits" (see also Goni et al. 2011). Pelc et al. (2010, p. 18266) stated that "Whereas the conservation benefits of marine reserves are clear, continuing debate about whether reserves can also benefit fisheries remains a major roadblock to successful implementation of reserves worldwide" (see also Gaines et al. 2010a). Ruttenberg et al. (2013, p. 116) claimed it all depended on the species: "For some species, trajectories were strongly influenced by MPA protection, but others showed no response to MPA implementation". Mora (2011) asserted that MRs can only protect about a fifth of marine fish stocks.

Christie (2011, p. 177) made the striking observation that the NP view of MRs tends to come from the North, whereas the SC view of MRs tends to come from the South: "It is notable that most of the…proponents for ambitious global MPA systems are…scientists and conservationists from the global North…[whereas] the academic community in the South is less comfortable with such large-scale policy agendas and favours priorities that elevate the needs of impoverished people".

Jones (2014, p. 42, 47) pointed out that "all the scientific consensus statements calling for MPAs have specifically been for the no-take rather than partially protected categories", but that the challenge to NTZs is growing, undermining "the previous apparent 'consensus' amongst the scientific community on the importance of no-take MPAs". The fact is that the MR case which was piloted so effectively by NPs from the 1990s, is now being challenged by a SC reaction. Although large scale MRs such as the Chagos Archipelago MPA (2010) have been designated in recent years, in other areas, such as Australia, proposed MR designations have been withdrawn: In 2013, the new centre right coalition government in Australia scrapped plans for 33 new MPA sites which would have expanded the country's MPA network from 27 to 60 sites, covering 3 million km^2, including the 1 million km^2 Coral Sea marine reserve, of which half would have been a no-take zone NTZ. "Environment Minister Greg Hunt told MPA News…that the new reserves were 'imposed without fair or adequate consultation' of industry, and would unfairly lock out recreational fishermen from large areas of the Ocean" (MPA News 15(3) Nov/Dec 2013, p. 1).

1.3 Growth of MPAs

The number of designated MPAs rose from 120 in 1970 to 10,280 in 2013 (Spalding et al. 2013; Devillers et al. 2014), and Spalding et al. (2013) pointed out that between 2003 and 2013 there was a fivefold rise in the area of MPA designations. This rise has occurred throughout the globe, but most quickly in the temperate Northern Atlantic region where the coverage increased from 1.6 to 12.9%. There has also been a rise in the designation of very large sites, including South Georgia (Antarctic), Chagos (Indian Ocean), Motu Motiro Hiva (Chile), Papahānaumokuākea Marine National Monument (Hawaii), Phoenix Islands (Kiribati), Pacific Remote Islands (USA), and Coral Sea (New Caledonia). If other areas were included in MPA designations—such as tracts of the Southern Ocean managed by the Convention for the Conservation of Antarctic Marine Living Resources (CCAMLR) (totalling 35 million km^2); the Nauru Agreement Concerning Cooperation in the Management of Fisheries of Common Interest which controls tuna fishing to prevent inter alia by-catch of whale sharks (totalling 4.5 million km^2); and the Indian and Southern Ocean Whale Sanctuaries—the total area of global marine protection would comfortably exceed the goal of 10% set by the CBD. Moreover, President Barack Obama announced in June 2014 that the USA was planning to create the world's biggest MR in the south-central Pacific by extending the Pacific Remote Islands National Monument (designated by President George Bush in 2009) to the 200 mile limit of US economic control around the islands (*The Guardian* 24.6.14). The UK is considering a proposal endorsed by the islanders to turn the waters surrounding the Pitcairn Islands in the middle of the Pacific Ocean into 836,000 km^2 MR (BBC News 26.6.14). As we shall see in Chap. 2, this huge rise in the number of MPAs reflects the endorsement of international agencies, treaties, environmental non-governmental organisations (ENGOs), and much of the marine science community. However, we should note that the vast majority (in numbers) of the designated MPAs are still MUMPAs rather than no-take MRs. For example, 85% of MPAs in the USA are MUMPAs, and MR areas occupy only 3% of US waters, most of which is in the Papahānaumokuākea Marine National Monument in Hawaii (NOAA 2013; Wenzel et al. 2013).

The growth of MPAs raises new issues of governance in relation to transboundary MPAs and high seas MPAs. As Guerreiro et al. (2010) pointed out, the location of species, habitats and ecosystems does not synchronise with political boundaries, and so states need to create joint mechanisms to protect biodiversity. On "transboundary MPAs" (TBMPAs) the protected areas that span the exclusive economic zones (EEZs) of two or more countries, the most obvious example is the EU: Although the vast majority of MPAs are located in the territorial waters of the member states, the European Commission has ruled that Natura 2000 requires an extension of the MPA networks to EEZs, and that member states coordinate their national MPA networks to form TBMPAs (e.g. the Dogger Bank Special Area of Conservation (SAC). Another example is in North America, where the North America Marine Protected Area Network (NAMPAN) comprises the USA, Canada, and Mexico, and affects a huge number of stakeholders along the Pacific coast (Guerreiro et al. 2010). A further example is in east Africa, where South Africa, Mozambique, and Tanzania are discussing the possibility of establishing a transboundary MPA in their common border areas (Guerreiro et al. 2010). However, Guerreiro et al. (2010) are not optimistic about the future prospects of TBMPAs, because the complexities of reaching agreement are immense, and for many regions, TBMPA will remain only a theoretical aspiration.

Despite the fact that the high seas comprise nearly 50 % of the Earth's surface and contain almost 90 % of its total biomass, they are the least protected habitats on earth (Hobday et al. 2011). Agardy et al. (2011) saw a bias against high seas (HS) MPAs, which Hobday et al. (2011) explained was because high seas MPAs (HSMPAs) were perceived to be impractical: Pelagic species swim vast distances and are impossible to protect over their entire range of travel; HSMPAs would be too large to police; little is known of the ecology of the high seas; many countries refuse to agree to restrictions on their fishing vessels in high sea waters; and there is no international body with authority to designate MPAs or regulate access to them (on the last point see Kim 2013; Molenaar and Oude Elferink 2009). Matz-Luck and Fuchs (2014) reported that the international law of the sea did not authorise states to take action against foreign vessels on the high seas for environmental protection reasons, but only against their own ships. Leonardo DeCaprio, the actor, called for HSMPAs in order to exert more control over fishing on the high seas: "Unfortunately today, there's no proper law enforcement capacity and little accountability for violating the law. It's the Wild West on the high seas. The ocean is an under-regulated marketplace…Even though the ocean covers 71 per cent of our planet, less than one percent are fully protected as marine reserves where fishing is prohibited" (*Fish Site* 20.6.14). David Miliband, former UK foreign secretary and now co-chair of the Global Ocean Commission (GOC), described the high seas as "virgin territory 35 or 40 years ago", but now "plundered territory" and the world's biggest "failed state" (*The Guardian* 24.6.14). White and Costello (2014, p. 1) claimed that "completely closing the HS to fishing would simultaneously give rise to large gains in fisheries profit…and fish stock conservation".

Nevertheless, there are already some HSMPAs. The first one was designated under the Oslo Paris Convention for the Protection of the Marine Environment of the North East Atlantic (OSPAR)'s auspices by Portugal in 2006—the so-called "Rainbow", which is located on Portugal's continental shelf outside its EEZ, administered by Portugal—an unprecedented extension of individual state authority on the high seas (Ribeiro 2010). In 2008, the World Wide Fund for Nature (WWF) reported four representative high seas MPAs where research was taking place, while FAO (2009) noted that some regional fisheries management organisations (RFMOs) had already closed to fishing some (very limited) areas of the high seas in the Mediterranean, Northeast Atlantic and Southern Ocean (Kim 2013; GOC 2013), though these designations only obliged states who were members of the organisations. In 2009, CCAMLR and the Antarctic Treaty designated the South Orkneys MPA—the world's first completely high seas no-take MPA—covering 94,000 km^2 of the Southern Ocean (GOC 2013; WWF 2010)—to protect foraging areas for penguins and other pelagic predators. In 2010, OSPAR established the first network of six HSMPAs, adding a seventh 2 years later, covering a total area of 470,000 km^2 in the Northeast Atlantic, protecting a series of seamounts and deep sea species and habitats (O'Leary et al. 2012). However, since OSPAR is a regional seas convention, unlike CCAMLR and the Antarctic Treaty, it has no legal power to regulate fisheries, and so its designated HSMPAs will have to be approved by the North East Atlantic Fisheries Commission (NEAFC), the International Commission for the Conservation of Atlantic Tunas (ICCAT), the North Atlantic Marine Mammal Commission (NAMMCO), the International Whaling Commission (IWC), and the International Maritime Organisation (IMO) (GOC 2013; Olsen et al. 2013; Molenaar and Oude Elderink 2009). Moreover, it is unclear which marine uses are still allowed in the OSPAR/Antarctic Treaty MPAs—OSPAR's guidance refers to six examples: scientific research, cable laying, dumping, construction of installations, building of artificial islands, and deep sea tourism. Furthermore, the GOC (2013) pointed out that only 2.8% of the world's oceans were designated as MPAs (compared with 12.7% of terrestrial areas), and only 0.79% of those areas were located on the high seas, which meant that the biggest ecosystem in the world was the least protected.

In order to speed up the designations of HSMPAs, an intense debate is currently taking place in international forums including the UN General Assembly Working Group on Biodiversity Beyond National Jurisdiction, over a proposal to complement the two existing United Nations Convention on Law of the Sea (UNCLOS) agreements which protect the seabed (the Seabed Authority Agreement) and fisheries (the Fish Stocks Agreement) respectively with a third agreement to protect biodiversity on the high seas by establishing a global regime that would bind all states (Kim 2013; Molenaar and Oude Elferink 2009). Such a regime was proposed by Corrigan and Kershaw (2008), who noted that both the International Union for Conservation of Nature (IUCN) and Greenpeace supported the idea of an implementation agreement under UNCLOS to create a high seas governance framework to establish a global MPA network. The GOC (2014) also endorsed this proposal, calling for a high seas regeneration zone which would close the high seas to industrial fishing. WWF (2010, p. 37) urged countries "to work towards a UN regime ensuring the recognition of all areas designated as MPAs in Areas Beyond National Jurisdiction (ABNJ) by states or mandated regional organisations". Many conferences held during the 2000s called for designations of HSMPAs based on the EBA (Kim 2013; Ardron et al. 2008). It seems that an epistemic community/pro-MPA advocacy coalition has been formed to promote a global network of HSMPAs (Corrigan and Kershaw 2008).

1.4 Structure of the Book

In Chap. 2, we examine the factors that led to the extraordinary rise of the MR paradigm to the most prominent policy position in present-day marine management policy. In Chap. 3, we confirm the prominence of the MR model in the academic literature. In Chap. 4, we investigate the scientific credentials of the case for MRs. In Chap. 5, we consider whether the scientific advocacy of MRs went too far beyond the evidence, and introduced a pro-MR bias into the peer-review system. In Chap. 6, we test whether the NP or the SC argument succeeded in the designation of MCZs in the UK case. In Chap. 7, we discuss the wider implications of the debate between NR and SC, including whether they could be reconciled in practice if not in principle.

Chapter 2
The Rise and Rise of the Marine Reserves 'Bandwagon'

2.1 Introduction

In this chapter, we investigate how and why the nature protectionist (NP) paradigm of marine reserves (MRs) became so prominent in the scientific literature during the 1990s and 2000s. The fact that the NP paradigm of MRs became dominant is demonstrated in Chap. 3 and is not in much dispute: What is less clear is how and why it did so. We argue that the key to its extraordinary rise in popularity is threefold: (1) it benefitted from a widespread perception that conventional fisheries management (CFM) had failed to prevent declines in fish stocks and in marine biodiversity around the world, and that a radical new approach was needed; (2) it owed much of its momentum to an elite group of marine ecologists who formed themselves into an epistemic community dedicated to the idea of MRs; and (3) it was taken up with enthusiasm by the international environmental movement who saw it as a worthy cause to prioritise and developed an advocacy coalition to promote it. We have already discussed the first factor in Chap. 1. In the present chapter, we discuss the second and third factors. On the third factor, we note that the pro-MR advocacy coalition was belatedly challenged by an anti-MR advocacy coalition, which has eventually succeeded in slowing down the progress of the MR bandwagon.

2.2 Policy Networks

The second factor (epistemic community—EpC) and the third factor (advocacy coalition—AC) are both examples of policy networks. Policy networks are aggregations of people who interact because of their common adherence to a set of views that impels them to seek policy change. Table 2.1 lists the characteristics which differentiate the EpC policy network from the AC policy network. We shall examine these characteristics in detail in the next two sections.

A. Caveen et al., *The Controversy over Marine Protected Areas*,
SpringerBriefs in Environmental Science, DOI 10.1007/978-3-319-10957-2_2

Table 2.1 Characteristics of the two networks

	Epistemic community	Advocacy coalition
Membership	Scientists/experts, and senior bureaucrats	Scientists, bureaucrats, elected officials, lobbyists, grassroots activists, industry, wider civil society
What binds members together?	Common body of knowledge	Principled beliefs
Decision-making model	Consensus	Compromise
Science-policy model	Linear model	Deliberative model
How does policy change occur?	Integration of experts of the international regime into their respective national governments, holding those governments to account	Policy change reflects the influence of competing advocacy coalitions, and unless one coalition is overwhelmingly dominant, a policy compromise usually results
Influence of the scientist	Scientists are central to policy change; they analyse the problem and set the policy agenda	Scientists align themselves with their preferred interest groups and offer their expertise in policy debate
Examples	Mediterranean pollution control; control of chlorofluorocarbons (CFCs)	MPAs in California; tropical deforestation

2.2.1 *Epistemic Community*

According to Sundstrom (2000), the concept of epistemic community (EpC) is a way of making sense of the fact that hard-to-grasp decisions may move actual, although not necessarily formal, power from elected representatives to elites acquainted with the subject in a transnational setting. Peter Haas (1989) first coined the term 'epistemic community' to describe the emergence of some international environmental regimes. An important feature of such regimes, in addition to their embodiment of rules and norms (Krasner 1983), is that they facilitate international learning and produce convergent state policies (Haas 1989). The notion of an EpC has been used to explain the coordinated response of states to many collective action problems at both the regional level (e.g. pollution control in the Mediterranean) and the global level (e.g. the regulation of CFCs) (Haas 1989, 1991). On the latter, Haas convincingly emphasised the role of scientific learning in the success of the Montreal Protocol, though critics like Sarewitz (2004) and Pielke (2007) have suggested that the ozone story was less of controversy resolved by science than of positive feedback from convergent scientific, political, diplomatic, and technological trends (including the fact that the main commercial interest—DuPont—eventually aligned itself with the main objective of the policy, that of phasing out CFCs, after it had developed CFC alternatives).

At the heart of the EpC is a group of experts who form around consensual knowledge, and share a policy enterprise (the action that needs to be taken to resolve an issue; e.g. the regulation of a hazardous chemical). The EpC is a useful theory for explaining policy responses to highly technical international problems where official decision-makers are unfamiliar with the technical details, and thereby unable to define state interests and develop viable solutions (Haas 1992b). This opens the door for a group of motivated individuals who through their expert understanding of the problem area, technical credentials, and common policy enterprise can offer potential solutions. The members of the EpC who are initially responsible for bringing states together to negotiate the international regime often have sufficient influence within their own governments to introduce regulation to their own domestic policy agenda (Haas 1989). The EpC is a good demonstration of the so-called 'linear model' of science-policy interaction, in that science is its fundamental bedrock, bringing to light new environmental problems and helping decision-makers to grasp their underlying causes; EpCs set the policy agenda. It is also a top-down model—the EpC is an elite group of scientists who tell truth to government on the problem that exists and the measures needed to overcome it. However, EpCs have had mixed success in practice: For example, while the Montreal Protocol has been viewed by some as very successful in limiting CFC emissions, the Kyoto Protocol has failed to curb global carbon dioxide (CO_2) emissions.

Evidence of the existence of an EpC committed to the cause of MRs comes from the fact that in the processes of getting provisions for MRs written into international regimes and agreements, leading roles were taken by a group of like-minded individuals in the United Nations Environment Programme (UNEP), Food and Agriculture Organisation (FAO), International Union for Conservation of Nature (IUCN), marine scientists, and MR planners and managers (Kelleher and Kenchington 1991; Salm et al. 2000; IUCN 2008). This community was united in its recognition of the MR as the best approach to protect marine biodiversity, and aimed at establishing MR networks to systematically protect representative habitats across each of the major marine provinces (OSPAR 2003b; Toropova et al. 2010). The policy recommendations of this EpC have been extensive: a number of guidelines and best practices have been provided by academics, ENGOs, research consultancies, and individual governments for the planning, development, management and evaluation of such MPA networks (Pomeroy et al. 2004).

Until 1985, only about 430 MPAs had been created, mainly covering relatively small coastal areas (De Silva et al. 1986; Bjorklund 1974), and few of these were MRs, mostly established for the purpose of scientific research—for example, Leigh Island, New Zealand (1975), Las Cruces, Chile (1982), and Apo Island, Philippines (1982). However, during the 1990s, the EpC became increasingly influential on the direction of international marine policy, and in 1992 the IUCN's Fourth World Congress on National Parks recommended that a global system of MPAs representing all major biogeographic types and ecosystems should be established. Later that year, the UN's Earth Summit in Rio called on coastal states to maintain biological diversity and productivity of marine species and habitats through the establishment and management of protected areas through the CBD. The ratification of the

CBD in 1994 placed a duty on signatory states to encourage 'projects that promote the conservation and sustainable use of biological diversity of coastal and marine resources under threat' (CBD 1994: 34(k)) implying the establishment of MPAs. However, explicit provisions for MPAs were not made in the CBD until 2006, when a target was adopted that stated that 10% of each marine and coastal ecological region should be conserved in MPAs by 2010. In 1995, the IUCN elaborated the idea of creating a representative system of MPAs for each of the world's major coastal biogeographic regions, identifying priorities for both regional and national authorities for establishing new MPAs or for improving management in those which already existed but were poorly managed or not managed at all (Kelleher et al. 1995). In 2003, the 12 coastal European nations of the Oslo Paris (OSPAR) Commission agreed to set up an 'ecologically coherent' network of MPAs in the North-east Atlantic by 2010, though no definition of ecological coherence was provided (Ardron 2008a).

By 1995 there were, globally, more than 1300 subtidal MPAs with a median size of 1548 ha (Kelleher et al. 1995). Towards the late 1990s and early 2000s, the norms and worldview of the EpC became institutionalised in high level policy guidance and international agreements such as the World Summit on Sustainable Development (WSSD) in 2002. In addition to their role in conserving marine biodiversity, MPAs were becoming increasingly recognised as a way to help rebuild the productivity of the oceans (Kelleher 1999). The policy enterprise of the EpC set the precedent for a pattern of decision-making that is largely top-down, with conservation goals for the planning area decided by experts. Some scholars argue that this has led to the dominance of the worldview of marine ecologists in the planning process (Christie 2011). Blount and Pitchon (2007) pointed out that in 1995, the World Bank, the Great Barrier Reef Marine Protected Area (GBRMPA), and the IUCN published a four-volume work on a global representative system of MPAs; in 1999, the University of Washington initiated *MPA News,* a monthly newsletter reporting developments in the theory and practice of MPAs worldwide; in 2001, the US Natural Research Council published a comprehensive history of MPAs; and in 2000, the US government established a national MPA centre to maintain a website database and library of publications on MPAs. North America Marine Protected Area Network (NAMPAN) was founded, and the first International Marine Protected Areas Congress was held in Geelong, Victoria, Australia in 2005.

Christie (2011, p. 177) stated that the 'MPA advocacy epistemic community has become remarkably influential and created a well-defined policy agenda which has diffused through influential conservation, resource management and donor institutions (and some government institutions)'. An illustration of the rising importance of the MPA EpC was the appointment of Dr Jane Lubchenco—an MR enthusiast—to be head of the US National Oceanographic and Atmospheric Administration (NOAA). As members of the MPA, EpC became integrated into advisory committees in their own governments, the domestic policies of these countries began to reflect the policies of the initial group of experts, for example through designing networks of MPAs to conserve nature by adopting ecological criteria (Airame et al. 2003; Roberts et al. 2003a, b). This was exemplified in England through the adoption of Natural England's (England's statutory conservation agency) Ecological Network Guidance (ENG) (Ashworth et al. 2010; see Chap. 6). According to Jones (2012, p. 249), the debate had moved on from 'whether MPAs are needed to how many MPAs are required, where they should be and how to design MPA networks'. By 2013 there were more than 10,000 MPAs in existence worldwide (Spalding et al. 2013).

However, behind the scientific credentials of the EpC's case for MRs to protect marine biodiversity lie several normative roots (Christie 2011). One normative root is the moral imperative of protecting species for their intrinsic value, an imperative that glossed over the trade-off between the protection of biodiversity and maintaining or increasing food supplies from the sea (Peterson and Lipcius 2003; Brander 2010). There were some divisions of opinion between scientists on the normative issue of the intrinsic value of biodiversity, and a significant fraction of the scientific community began to frame the empirical debate around what MRs could achieve for fisheries (Roberts 1997; NCEAS 2001), with the emergence of the American consensus statement on the fisheries benefits of MRs published in 2001 (NCEAS 2001). Another normative root was the scientists' understanding of how concepts such as biodiversity, ecosystem resilience and ecosystem productivity should be interpreted, which raised the questions of what value an ecosystem state has over another; who/what benefits from protecting an ecosystem in a certain state by imposing an MPA on it; and how we are to calculate the value of ecosystem's goods and services. A further source of division within the EpC opened up over the scientific justification of percentage targets for MRs based on the findings of abstract modelling studies. Several researchers tried to prove through modelling exercises that 10–50 % (modal value of 30 %) of the oceans should be designated as MRs to sustain fisheries. Roberts claimed that 'The consensus of scientists is that 30 % of our oceans should be dedicated fishing exclusion zones' (Hastings 2009). However, Agardy et al. (2003) pointed out that this 'rule of thumb' originated from modelling studies that were principally focused on tropical coral reefs (Bohnsack et al. 2000), which did not offer much support to the scientific case for MRs as a preferred tool in the management of temperate marine resources. Jones (2006, p. 144) alluded to further divisions between scientists over the effectiveness of MRs in delivering fish stock protection, and concluded that 'the consensus on the need for NTZs may actually be confined to quite a narrow constituency'. It may be, therefore, that while MR advocates have aspired to form an EpC, they are more aptly described as an AC, using science to justify their differing value preferences.

2.2.2 Advocacy Coalitions

Epistemic communities are successful when their core policy enterprise remains unchallenged at all levels of government and little significant opposition within the scientific community exists to refute their causal and normative assumptions. They are less successful where the problem area is more opaque in its causal underpinnings or where irreconcilable differences exist in actors' fundamental normative beliefs. In such cases, the AC (Sabatier 1988; Keck and Sikkink 1998)[1] is likely to be more effective (Sabatier 1998; Weible and Sabatier 2005). Unlike EpCs, ACs are not limited to 'knowledge experts' (i.e. academics and public servants), but also include non-governmental organisations (NGOs), think tanks, journalists, celebrities and members of civil society. Whereas the norms of the EpC manifest themselves in a 'regime' that tries to impose its rules and regulations on its members, the norms of the AC are manifested in a less formal 'cluster' of people with a common cause. Nevertheless, actors belonging to the AC are bound together by common values, extensive exchanges of information and services, and a shared discourse (Stone 2002), and the coordinated action of all these actors can be a powerful stimulus to policy change.

The origin of the MR advocacy coalition can be traced back to the late 1970s when Bill Ballantine (widely regarded as the founding father of marine reserves) campaigned for several years to establish New Zealand's first MR at Leigh (see Table 3.1) and continued to lobby the New Zealand government until more were established. The New Zealand experience is often cited by advocates of MRs as the ideal to be achieved and has captured the attention of high profile journalists (e.g. Clover 2004). The campaign efforts of this AC also became more transnational in scope. For instance, in the late 1990s, there was a growing advocacy literature for MRs from scientists (e.g. Roberts 1997; Lauck et al. 1998; see Sect. 2.3.4), including Jane Lubchenco (later head of NOAA) who in 1997 called for the protection of 20 % of the ocean as no-fishing zones by 2020. In 1998, the US Marine Conservation Biology Institute (MCBI) issued *Troubled Waters: A Call for Action,* a statement signed by 1605 scientists across the globe bringing to the world's attention the damage being caused to the oceans. One of its five recommendations was to increase the number and effectiveness of MRs so that 20 % of EEZs and the high seas were protected from threats by the year 2020 (restating Lubchenco's plea 1 year previously). Another recommendation was to restrict or stop fishing methods that undermined sustainability by harming the habitats of economically valuable marine species and the species they used for food and shelter. This statement coincided with a highly influential research paper funded by the MCBI that documented the global impact of bottom trawling on seabed habitats (Watling and Norse 1998).

Another significant element of the MR advocacy coalition during the later 2000s was the popularisation of the MR idea within wider society, and the key role that scientists took in this popularisation. For example, the marine conservation scientist Callum Roberts devoted a chapter to MRs in his highly influential[2] book *The Unnatural History of the Sea*, in which he argued that marine reserves must be

[1] The term 'advocacy coalition' has been used by Sabatier (1988) and other authors to explain the actions of advocacy networks operating solely at the domestic level. However, we use the term to explain the actions of international advocacy groups that are named elsewhere as 'transnational advocacy networks' (TANs; Keck and Sikkink 1998).

[2] An illustration of the book's influence is that it was mentioned during the House of Lords debates on the UK Marine and Coastal Access Act.

extensive, covering between 20 and 40 per cent of the sea, in order to sustain ecological processes and services—like fisheries—that are vital to humanity' (Roberts 2007b, p. 382). The MCBI along with the Pew environmental group in 2006 and 2009 persuaded President George Bush Jr. to establish three large MRs in the tropical Pacific Ocean. The Pew Trust also championed the MR protection of the Chagos archipelago (for which the Blue Marine Foundation[3], a UK-based environmental group, secured financing for the first 5 years; Roberts 2012), Australia's Coral Sea, and New Zealand's Kermadec Islands.

Another difference between EpC and AC is that there may be several ACs within the policy community competing to get their voices recognised by government, and policy change may be a result of shifts in power between competing ACs (Schlager 1995). The relationship between knowledge and power in such an AC configuration is an example of 'interest group pluralism' whereby scientists best serve society by aligning themselves with their favoured faction or interest group, offering their expertise as an asset in political engagement (Stone 2002; Pielke 2007). Evidence of the existence of one AC *for* MRs, and another AC *against* MRs, can be found at the international level. The pro-MR AC exists in the highly coordinated network that has developed within the global environmental movement (e.g. IUCN, World Wildlife Fund—WWF, Pew Trust). It has had especial impact through the Pew Environmental Trust's global ocean legacy scheme which aims to establish a worldwide system of very large (300,000 km^2) MRs, four of which have now been established. Another powerful influence on framing MPA policy debates is the Partnership for Interdisciplinary Studies of Coastal Oceans (PISCO), which has run a campaign for MRs, producing a series of educational booklets showing their positive ecological effects (PISCO 2011). Hilborn et al. (2004, p. 198) referred to this pro-MR advocacy coalition as 'Globally…a wave of environmental groups, politicians and ecologists pushing for the large-scale implementation of MPAs, with many calls for protecting 20–30 % of the oceans'.

Pajaro et al. (2010) described the pro–MR AC as a transnational advocacy network (TAN), which originated in the 1960s when many social networks which shared common ideas about the deficiencies of society were formed to lobby policymakers. Examples of TANs that espouse the MR cause include the IUCN, the Global Coral Reef Monitoring Network, and the International Coral Reef Action Network. 'TANs have a tendency to be dominated by international NGOs with established ties to the UN system and thus tend to have better access to resources for networking and lobbying when compared to national or local MPA organisations' (Pajaro et al. 2010, p. 948).

The anti-MR AC exists in the more recent network formed by the global fishing industry manifested, for example, in the International Coalition of Fisheries Associations which assembled in November 2007 to identify and address issues of common interest in international fisheries and called on their governments to recognise the limitation of MRs as a fish stock protection measure. This challenging anti-MR AC reflected growing scepticism among some scientists regarding the

[3] This group was initially conceived by the people who were behind the film *The End of the Line*.

value of MRs for fisheries (Hilborn et al. 2004; Kaiser 2005; Jones 2007). Worries about advocacy for MRs highlighted six concerns: (1) that MRs would not meet their objectives unless scientists had a good understanding of the local ecological and socio-economic context (Christie 2004); (2) that the assumption that MRs would bring more benefits than MPAs made stakeholders suspicious that scientists had normative motivations for establishing them (Jones 2002; Agardy et al. 2003); (3) that claims for MRs may be based on overgeneralised science regarding their fisheries effects (Hilborn et al. 2004; Kaiser 2005); (4) that claims for MRs may not be backed by robust empirical evidence (Willis et al. 2003a); (5) that alternative management options may be more appropriate to achieve certain objectives (Steele and Hoagland 2004; McClanahan 2011; Jones 2007); and (6) that targets for percentages of MRs in global oceans were counterproductive, focusing too heavily on means (percentage of area covered by MPAs) rather than ends (improved conservation outcomes) (Pressey 2013).

The pro- and anti-MR AC framework has been used to explain the decision-making process during the establishment of a network of MRs in California (Weible 2007)[4]. In England, both pro- and anti-MR ACs have had influence on government during the planning of the MCZ network (see Chap. 6). The pro-MR AC was represented by environmentalists (including the Marine Conservation Society—MCS, the Marine Network of Friends of the Earth Local Groups (Marinet), the Wildlife Trusts, WWF, Royal Society for the Protection of Birds (RSPB), Natural England (NE), and committed scientists), while the anti-MR AC was represented by the fishing industry, centred mainly on the MPA Fishing Coalition (MPAC), headed by Dr Stephen Lockwood, an ex-CEFAS (Centre for Environment, Fisheries and Aquaculture Science) fisheries scientist. From a science-policy perspective, the pro-MR AC emphasised the scientific credentials of MRs, while the anti-MR AC criticised the policy recommendations of the environmentalists, pointing to the adverse ecological impacts and socioeconomic costs of the displacement of fishing effort after an MR is established (see Chap. 6). In both ACs, science was viewed by their proponents as a resource for enhancing their ability to advocate, bargain, and negotiate in pursuit of their special interests.

[4] It is important to state that the words 'pro-MR' and 'anti-MR' are shorthand terms for, respectively, 'favouring widespread networks of no-take MPAs across the oceans' (pro-MR); and 'opposing the unselective establishment of MRs' (anti-MR). The anti-MR AC is not opposed to all MRs, but only to those that have insufficient scientific and/or socio-economic justification. Indeed, many members of the anti-MR AC are enthusiastic supporters of MRs in the right places, and would be classified as 'social conservationists' (see Chap. 1). As we will see in Chap. 5, the anti-MR coalition in the English case fully recognised the value of MPAs in protecting vulnerable species and seabed features: 'The issue is not whether there should or should not be MPAs. It is about a rational, fair and balanced process in their establishment…against a sometimes irrational push by naive enthusiasts, who see MPAs as an all-embracing solution for overfishing—and all the other ills of the marine environment' (*Fishing News* 28.2.14, p. 5).

2.3 Conclusion

This chapter has traced the progress of the MR 'bandwagon', showing how it developed its overwhelming momentum through the efforts of marine ecologists forming themselves into an EpC committed to the establishment of a global network of MRs, and exerting considerable pressure on policy makers. The MR cause was also taken up enthusiastically by international ENGOs and other environmentalists who formed an AC to promote it. However, the opponents of MRs belatedly responded to these pro-MR networks and established an anti-MR AC to challenge what appeared to be an inexorable move to impose MRs across the world's oceans, and this challenge seems to have slowed down the MR juggernaut, at least temporarily.

Chapter 3
Bibliometric Test of the MR 'Bandwagon'

3.1 Introduction

This chapter presents a bibliometric analysis which is designed to test the claims made in Chap. 2 of a marine reserve (MR) 'bandwagon'. This bibliometric analysis of the peer-reviewed MR literature made use of a social network analysis to identify key scientists and a citation analysis to identify key papers. The social network analysis was performed to find out which scientists were most connected with their peers through research collaboration. The citation analysis was designed to discover which MR papers have been most cited, in which journals they have been most published, and the extent to which there was crossover in the most highly cited papers between different research fields. The findings of the bibliometric analysis are that MR studies dominated the marine management literature, that 90 % of MR researchers are marine ecologists, that MR publications have been highly influential among marine scientists, and that MR research has likely attracted more funding than any other subject area in the applied marine sciences. This is an emphatic confirmation of the MR 'bandwagon'.

3.2 Methods

Keywords including marine-protected areas, marine reserves, marine parks, marine sanctuaries, no-take marine zones, special areas of marine conservation, and closed marine areas were employed to identify all literature published on marine-protected areas (MPAs) between 1970 and 2010 from ISI's Web of Science (WoS). Records were imported into HistCite^{TM}, and authors were ranked according to their publication count. Data obtained from the search included empirical, theoretical, natural, and social science articles, as well as opinion pieces and reviews.

A. Caveen et al., *The Controversy over Marine Protected Areas*,
SpringerBriefs in Environmental Science, DOI 10.1007/978-3-319-10957-2_3

3.2.1 Social Network Analysis

Social network analysis views social relationships in terms of network theory, and can be used to study a wide range of social phenomena such as friendships and communication networks (De Nooy et al. 2005). It allows the measurement of an actors' centrality (an indication of prestige) in a social network, which, in the present investigation means the position of leading scientists in a coauthor network that has studied MRs. Authors selected for study as a first network had to satisfy the following conditions: (1) a minimum of 10 peer-reviewed publications on MRs, (2) a minimum local citation score (LCS)[1] of 100, and (3) connection to at least one other author. A total of 52 authors met these conditions, and constituted a network which was laid out using the Pajek 1.24 (Batagelj and Mrvar 2004) network visualisation package. This coauthor network was analysed in three timescales—1970–2000, 2001–2005, and 2006–2010—to trace the pattern of when collaborations evolved. A second network was located in relation to the scientists who signed the North American (NCEAS 2001), European (Roberts 2007a), and Australian (AMSA 2008) consensus statements, to test for a link between membership of the first network and explicit involvement in advocacy for MRs. Two measures of a scientist's centrality in a network were used—'closeness' and 'betweenness' (Batagelj and Mrvar 2004). Closeness measures the total distance between one vertex and all other vertices, with larger distances yielding lower closeness centrality scores. Betweenness measures the extent to which an actor is an intermediary between pairs of vertices, or, put another way, the number of times a node needs a given node to reach another node.

3.2.2 Citation Analysis

Citations are increasingly being used as a metric to evaluate a scientist's/research group's/institution's impact (Andras 2011). By citing a paper, authors of an article are implying that in conducting their research, they have been influenced by the cited source. Two assumptions of this study are that more highly cited papers have had greater impact in (a) shaping scientific thinking on a subject (Marx and Bornmann 2013) and (b) influencing policy debate due to the high amount of publicity such articles often attract through the mass media (Hilborn 2007b) and popular literature (Clover 2004; Roberts 2007b). These are controversial assumptions, in that an article might be highly cited but have little impact on either scientific thinking or policy debate. However, the balance of probabilities is that highly cited papers are more influential than are non-cited papers.

To discover paper-citation networks, the same search term and time period as for the construction of the coauthor networks was used, and the top 20 most highly

[1] The LCS score is a measure of a scientist's influence within his/her 'local' (defined by the search terms used to find publications in WoS) science community.

Table 3.1 A summary of the number of publications for different search terms (1970–2010) and percentage overlap between search X with searches N_{1-4}. Global citation score (GCS) is the total number of times an article has been cited

Database	Search term used to source data from WoS	Number of records (end 2010)	Percentage of shared records with search X	Total GCS of database	Mean GCS per article
X	Marine and ('marine reserve'...)	5294	n/a	79,059	14.9
N_1	Marine and fisher	6691	22.83 %	86,384	12.9
N_2	Marine and conservation	3784	29.92 %	58,452	15.4
N_3	Marine and management	8077	24.65 %	90,959	11.3
N_4	Marine and policy	1143	17.13 %	12,888	11.3

WoS Web of Science, *GCS* global citation score

and locally cited papers contained within the database were identified. To find out how prominent MPAs had become in the wider marine literature, the proportion of literature on MPAs that existed in other research fields was calculated by means of a comparative analysis of the original search term with seven additional search terms (see Table 3.1).

3.3 Results and Discussion

It is important to identify the members of the scientific community who had published most extensively on MRs up until the end of the twentieth century. The structure of the coauthor network shows two sub-networks consisting of scientists from the USA and Europe (see Fig. 3.1), members of the former being particularly influential in the numbers of both publications and citations (Caveen et al. 2013). Influential institutions include National Oceanographic and Atmospheric Administration (NOAA), James Cook University, and the University of Auckland. Regarding the academic impact of individual scientists, TR McClanahan has the greatest number of publications; CM Roberts has the greatest LCS; and NVC Polunin has the greatest centrality scores for both closeness and betweenness. The network consisted mainly of scientists who would best be described as marine ecologists. Only two social scientists, A White (The Nature Conservancy, USA) and P Christie (University of Washington, USA) whose work focused more on the social aspects of marine resource management, had sufficient publications to be included in it.

With respect to the dissemination of knowledge, the most published-in journals included the *Bulletin of Marine Science* (which published a special issue on MPAs in 2000 on their role in the protection of essential fish habitat), and *Marine Ecology*

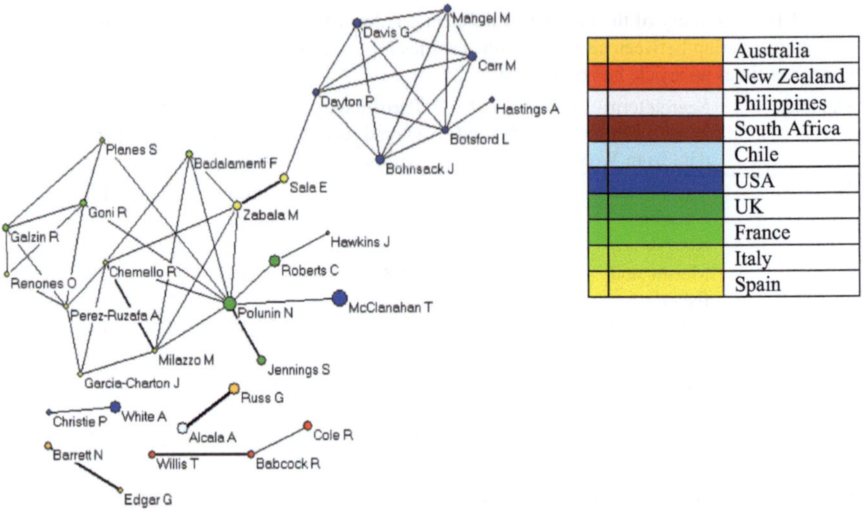

Fig. 3.1 Coauthor network of the most productive authors in MR science from 1970 to 2000 ($n=32$). *Vertex size* indicates the relative number of publications for an author and *edge width* the number of times an author has collaborated. *Vertex colour* indicates author nationality

Progress Series (MEPS). The most studied MPAs were all MRs, mainly located over tropical coral reefs (Leigh, NE New Zealand being the exception). Importantly, such studies are amongst the most highly cited in the MPA literature (see Fig. 3.2). It is reasonable to assume that these highly cited studies have influenced public discourse on MPAs (see Chap. 1), and the subsequent framing of policy debates—particularly the perceived need to establish MRs (see Sect. 3.3.2).

However, the networks shown in Figs. 3.1 and 3.2 overlook scientists who have spent more time advising policymakers than writing academic papers, and influential pieces of 'grey' literature that may not get formally published. An example is Bill Ballantine's (New Zealand) keynote address at 'The Design of and Monitoring of Marine Reserves Workshop' held at the University of British Columbia's Fisheries Centre in 1997 (Ballantine 1997). The address was not published but it does lay the foundations for principles that govern the design of MR networks. Ballantine's design principles were subsequently elaborated by a National Center for Ecological Analysis and Synthesis (NCEAS) working group on the design of MR networks in 1999 (also see next section), the output of which was published in a special issue of *Ecological Applications* in 2003, and these papers became highly cited in both academic and policy documents (e.g. Roberts et al. 2003b; Ashworth et al. 2010).

From the early 2000s, the scientific community became more connected (see Figs. 3.3 and 3.4 for the evolution of coauthor networks for 2000–2010). Scientists from the USA continued to be influential with a combined 666 publications and an LCS of 2333. Eight of the most published authors were US nationals, many of whom received funding through NCEAS. The closeness scores for many scientists had increased; CM Roberts being the most central actor in the network (0.373). Both NVC Polunin and CM Roberts were key intermediaries (information brokers) in the network (with scores for betweenness of 0.278 and 0.233, respectively).

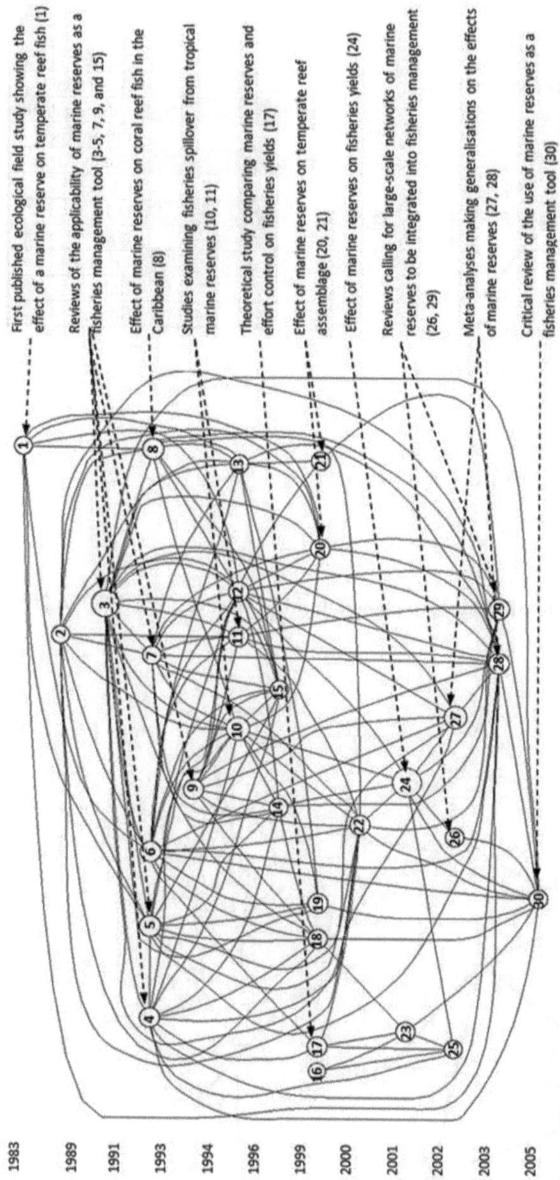

ID	Paper	LCS	GCS	Type
1	Bell JD, 1983, J Appl Ecol, V20, P357	95	132	E, Te
2	Russ, GR, 1989, Mar Ecol Progr Ser, V56, P13	92	139	E, Tr
3	Roberts CM, 1991, Reviews in Fish Biology, V1, P65	204	260	R, Tr
4	Carr MH, 1993, Can J Fish Aquat Sci, V50, P2019	113	144	R, Te
5	Dugan JE, 1993, Can J Fish Aquat Sci, V50, P2029	133	171	R, Te
6	DeMartini EE, 1993, Fish Bull, V91, P414	122	135	T, Tr
7	Roberts CM, 1993, Ambio, V22, P363	98	116	R, Tr
8	Polunin NVC, 1993, Mar Ecol Progr Ser, V100, P167	133	175	E, Tr
9	Rowley RJ, 1994, Aquat Conserv, V4, P233	117	133	R, Te
10	Russ GR, 1996, Mar Ecol Progr Ser, P947	156	168	E, Tr
11	McClanahan TR, 1996, Conserv Biol, V10, P1187	119	143	E, Tr
12	Russ GR, 1996, Ecol Appl, V6, P947	156	168	E, Tr
13	Rakitin A, 1996, Mar Ecol Progr Ser, P97	102	113	E, Tr
14	Bohnsack JA, 1998, Aust J Ecol, V23, P298	98	115	R, Tr
15	Guenette S, 1998, Rev Fish Biol Fisheries, V8, P251	92	108	R
16	Swearer SE, 1999, Nature, V402, P799	87	377	E, Tr
17	Hastings A, 1999, Science, V284, P1537	116	141	T
18	Murray SN, 1999, Fisheries, V24, P11	106	135	R
19	Kramer DL, 1999, Environ Biol Fish, V55, P65	113	171	E, Tr
20	Edgar GJ, 1999, J Exp Mar Biol Ecol, V242, P107	106	125	E, Te
21	Babcock RC, 1999, Mar Ecol Progr Ser, V189, P125	98	140	E, Te
22	McClanahan TR, 2000, Ecol Appl, V10, P1792	119	150	E, Tr
23	Botsford LW, 2001, Ecol Lett, V4, P144	103	155	T
24	Roberts CM, 2001, Science, V294, P1920	234	323	E, Tr
25	Sala E, 2002, Science, V298, P1991	95	141	T, Tr
26	Pauly D ,2002, Nature, V418, P689	102	614	R
27	Halpern BS, 2002, Ecol Lett, V5, P361	148	183	E
28	Halpern BS, 2003, Ecol Appl, V13, P117	121	182	E
29	Gell FR, 2003, Trend Ecol Evolut, V18, P448	123	177	R
30	Sale PF, 2005, Trend Ecol Evolut, V20, P74	102	197	R

Fig. 3.2 Most highly cited studies on MPAs. *LCS* local citation score, *GCS* global citation score

In 2000, a much-cited report from World Wide Fund for Nature (WWF) which showcased the fisheries effects of MRs (Roberts and Hawkins 2000), heralded a flurry of academic publications in the early 2000s that documented the potential of MRs to benefit fisheries, albeit with evidence limited to a few case studies (Roberts et al. 2001; Russ et al. 2003, 2004). This shaped the climate of thinking for the next

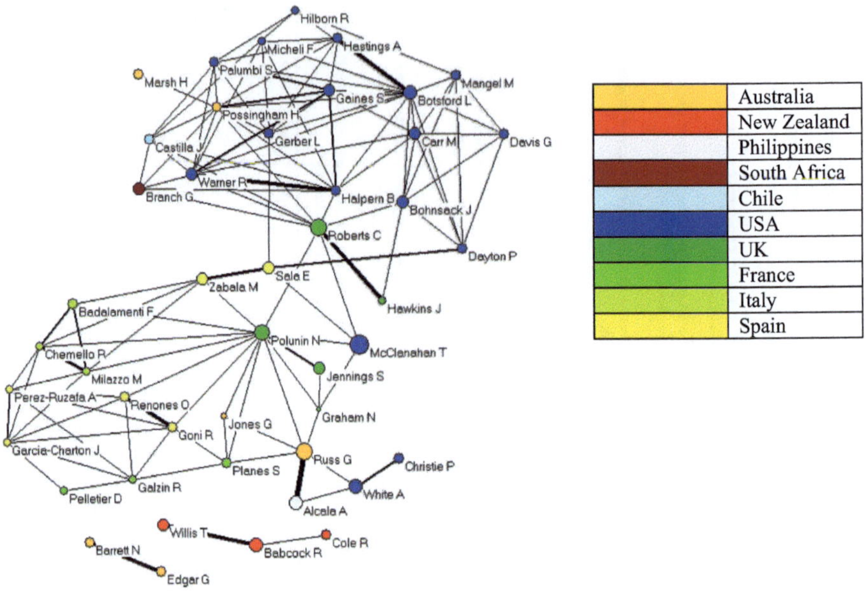

Fig. 3.3 Coauthor network of the most productive authors in MPA science from 1970 to 2005 (n=41)

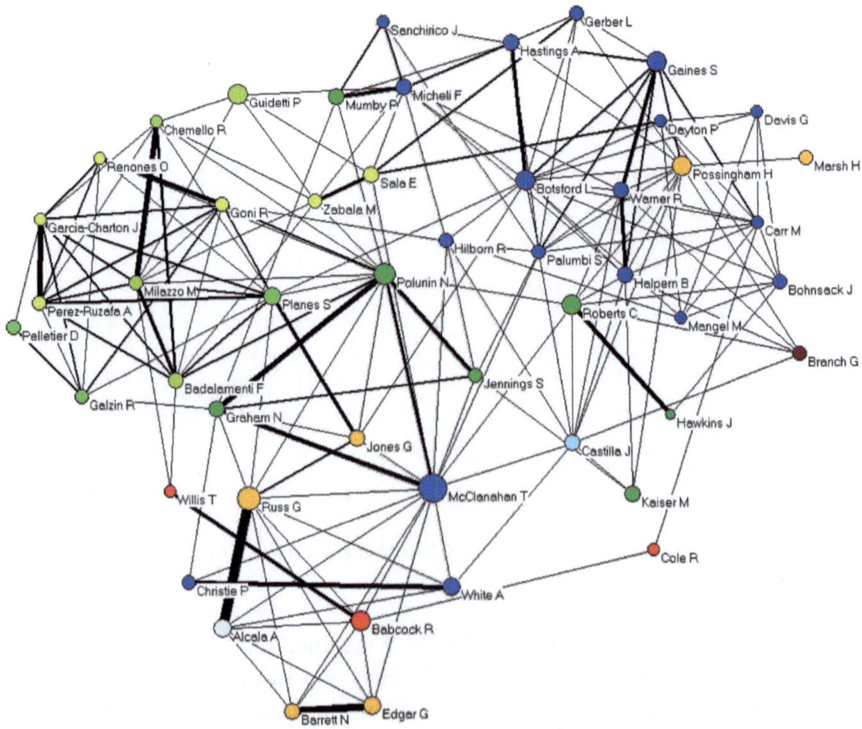

Fig. 3.4 Coauthor network of the most productive authors in MPA science from 1970 to 2010 (n=48)

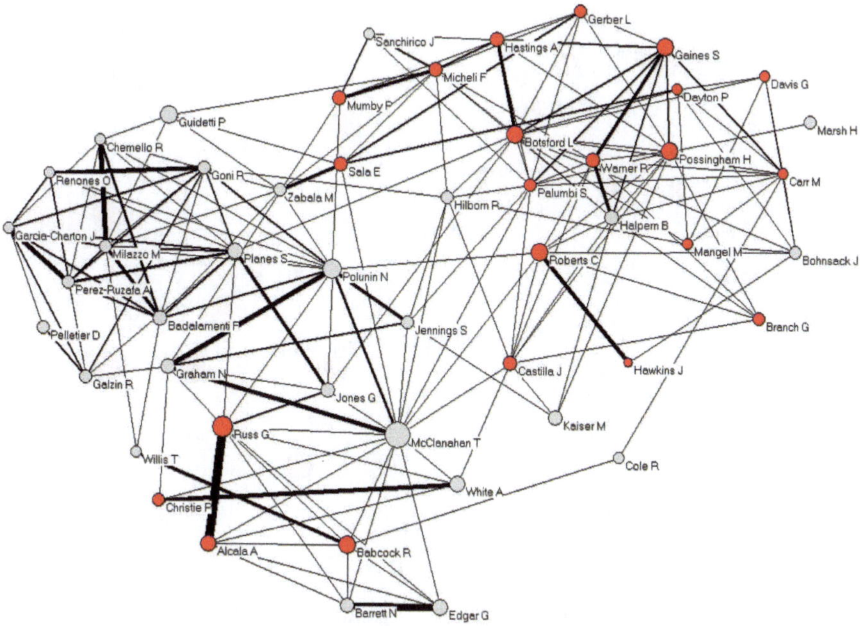

Fig. 3.5 Coauthor advocacy network of the most productive authors in MPA science from 1970 to 2010 ($n=48$) as in Fig. 3.4. *Vertices* are coloured *red* according to whether the scientist was a signatory to the North American, European, and Australian consensus statements on MRs. (NCEAS 2001; Roberts 2007a; AMSA 2008)

decade, as scientific debate became heavily focused on what MRs could do for subsistence and commercial fisheries rather than their wider potential benefits for the conservation of noncommercial species (Edgar 2011). The underlying political agenda of these highly cited studies was clear; Gell and Roberts (2003) argued that to reverse global fishery declines it was necessary to integrate large-scale networks of MRs into fisheries management. In his highly influential piece for *Nature,* Daniel Pauly implied that zoning of large areas of the oceans as MRs was essential if fisheries were to be sustainable (Pauly et al. 2002). Additionally, a metanalysis by Halpern and Warner (2002) showed that the responses in terms of density, biomass and mean size of fish and invertebrates to protection in MRs appear to occur quickly, which the authors believed should facilitate the use of MRs in the management of marine resources. This promoted a discourse that documented the global collapse of fisheries and implied that MRs were the primary tool required to avert such a catastrophe.

A substantial proportion of the MR scientific community became involved in some sort of advocacy for them (see Fig. 3.5). Based on the signatories to the three consensus statements, the strongest tendency towards advocacy for MRs came from US scientists. This corresponds well with the heavy involvement of US scientific institutions such as NCEAS in funding knowledge production on MRs (e.g. see

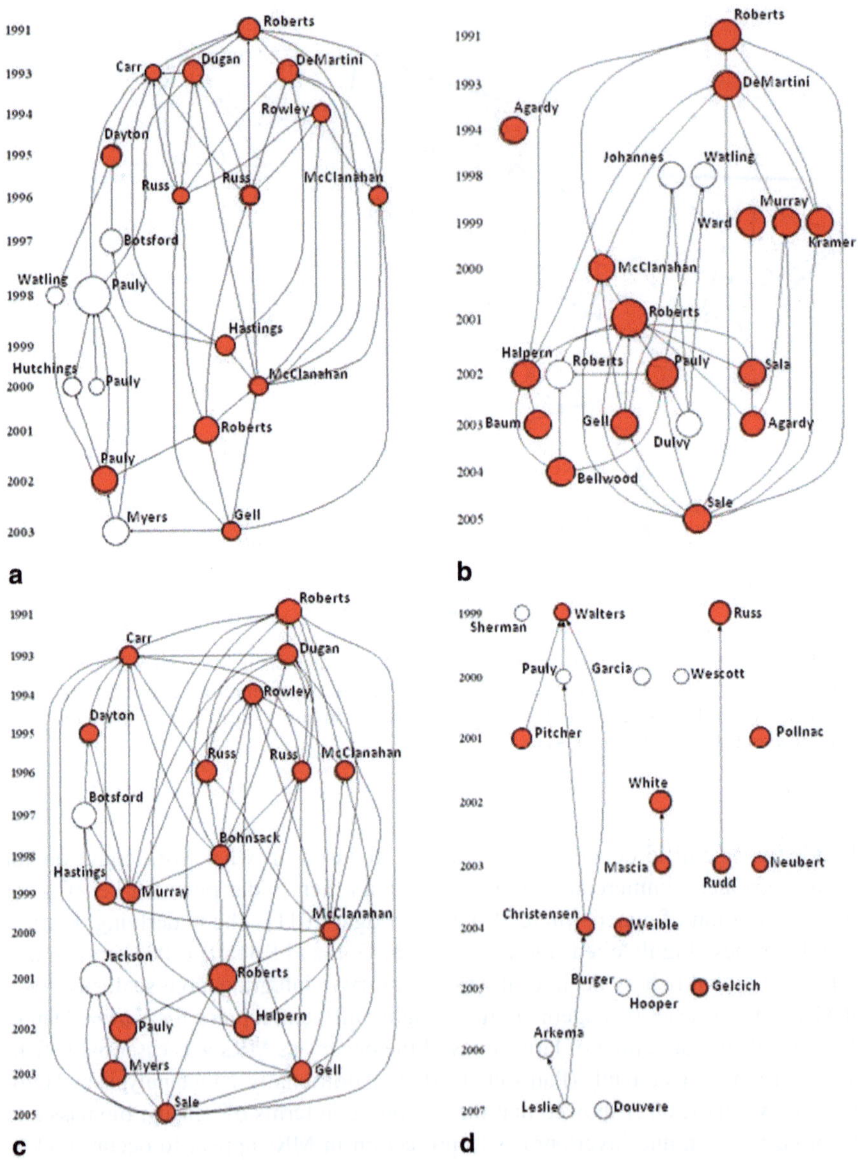

Fig. 3.6 Paper citation networks of the top 20 papers from searches N_{1-4}. **a** *Marine and fisher* (threshold 100 citations). **b** *Marine and conservation* (threshold 50 citations). **c** *Marine and management* (threshold 86 citations). **d** *Marine and policy* (threshold 6 citations). The *node size* denotes relative number of citations that a paper has in its respective database. *Shaded nodes* indicate papers that are also present in database X (i.e. *marine and 'marine reserve'*)

Ecological Applications volume 13, 2003), and Partnership for Interdisciplinary Studies of Coastal oceans (PISCO) in disseminating MR science (PISCO 2007, 2011). There may also have been a more accepting attitude among US-based conservation journals towards advocacy (Scott et al. 2007).[2]

The greater involvement of US scientists in advocacy for MRs may also stem from the more deeply rooted commitment by the US to marine protection (Sloan 2002), compared to countries in Europe that are in the process of establishing MPA networks (e.g. Natura 2000). For instance, the US Marine Sanctuaries Act (1972) provided the means to protect unique habitats, and the US Magnuson-Stevens Fishery Conservation and Management Act (1996) legislated for the use of MRs as a management tool to protect essential fish habitats. Another reason is institutional differences; the politics of fisheries management in the EU has meant that EU member states, through the adoption of the Common Fisheries Policy (CFP), have conceded to the EU the territorial use rights for their EEZs outside of the 12 nm limit. This has posed a collective action problem for EU fisheries management and conservation in establishing MR networks (Fock 2011). MR networks are more easily and successfully (in terms of compliance and performance) applied in countries that have full control over their EEZs (e.g. the USA, New Zealand, and Australia).

Since studies on MRs have dominated the marine management literature (see Fig. 3.6), there is a danger of diversion of research away from other areas of the science-policy agenda where investigation is also desirable. As Metcalfe (2013) noted, research into MRs has been much greater than research into multi-use MPAs (MUMPAs).

3.4 Conclusion

This chapter sought bibliometric evidence to test for the existence of a dominant MR paradigm in marine science. The findings of the social network analysis showed that MR studies dominated the wider marine management literature; that of the 48 scientists identified in the coauthor network, the majority (around 90 %) were marine ecologists, not fisheries scientists, suggesting that MRs are an integral part of the heuristic in which marine ecologists have been trained (Degnbol et al. 2006); and that a large number of highly cited MPA studies are also found in other research fields, implying that a much higher amount of funding has been spent on MPA and MR research than on other management interventions. The citation analysis showed that MPA publications have been highly influential within the marine science community as measured by the number of citations and by crossover with other research fields (see Fig. 3.6). The next chapter deals with the issue of whether this dominance by the MR paradigm has led to a bias in the peer-reviewed literature towards pro-MR studies, and if so, whether this means that the distinction between science and policy advocacy has become blurred.

[2] In 2012 Erica Fleishman, editor-in-chief of the Society for Conservation Biology's flagship journal *Conservation Biology*, was forced out by the board due to her insistence on removing advocacy statements from research papers.

Chapter 4
Bias in the Peer-reviewed Literature, and Crossing the Science/Policy Divide

4.1 Introduction

One important issue raised by the dominance of MR in marine science is whether there is bias in the way in which scientific research is published. Does MR prominence mean that papers which support the orthodoxy in favour of MRs are more likely to be published than are papers which question MRs? This is the focus of Chap. 4, in which there are three key questions: (1) is there any evidence to suggest that a pro-MR bias exists amongst scientists? (2) If so, does this bias affect the type of results published in the MPA literature? (3) If so, does this mean that the pro-MR camp has crossed the line between science and policy advocacy? A short questionnaire was sent to 200 leading scientists who have studied the ecological effects of MRs, to test for such a bias. The questionnaire focused on two issues: scientists' experience of having publications rejected; and scientists' attitudes towards publishing non-significant research findings. The results did not find evidence of a systematic pro-MR bias, but this does not necessarily mean that pro-MR scientists have not crossed the line between science and policy advocacy.

Although, some studies have addressed the question of whether there is bias in the peer-review publication of scientific knowledge (Fanelli 2012), no such study has focused on the MR issue. It is true that in the marine natural resources literature some authors have been critical of the peer-review process, arguing that it may not always be a guarantee of objectivity (Hilborn et al. 2004; Kaiser 2004; Banobi et al. 2011). The FAO (2006, p. 63, para 6) stated that 'Many of the negative media reports are based on presumably authoritative scientific publications that are assumed to have been adequately "quality assured" by journal peer review. However, journal peer reviews are conducted by mail by a few scientists, often with little first-hand knowledge of the fisheries being considered, and rarely with access to the data upon which the paper they are reviewing is based. This is only a cursory level of quality assurance'. It is also true that in MR science some authors have suggested that there is a publication bias in the MR literature in favour of studies that show positive ecological effects (Huntington 2011; Hilborn 2006). For example, Cressey (2011b, p. 167) quoted Peter Sale saying that 'There's almost a religion, which is nurtured

A. Caveen et al., *The Controversy over Marine Protected Areas*,
SpringerBriefs in Environmental Science, DOI 10.1007/978-3-319-10957-2_4

by non-governmental organisations and some members of the science community, to the extent that critical examination of MPAs is exceedingly difficult to get published', while Jones (2006, p. 148) pointed out that 'Willis et al. (2003) argue that the *raison d'étre* for many review and theory papers concerning NTZs is advocacy for such designations, rather than real attempts to contribute to the science of the field'.

However, no study has systematically investigated these claims and we aim to fill this gap by examining whether there is evidence to validate them. We ask whether there is a pro-MR bias in the peer-review process, and if so, whether it lies in the hands of journal editors and reviewers who reject studies that they think may interfere with the uptake of MRs as a management tool, or whether it lies in the hands of the scientific researchers themselves who choose not to publish certain research findings or frame their research questions in a way that precludes the discussion of non-significant and/or negative effects of MRs (i.e. self-censorship)? Of course, even if such self-censorship exists in MR research, it may reflect a more general problem in science (Fanelli 2012), the under-reporting of non-significant ($P>0.05$) results (the so-called 'file drawer problem') (Rosenthal 1979).

The question of bias has wide ramifications—including the possibility of crossing the science/policy advocacy divide. MR science is clearly issue-driven in that it is charged with finding the circumstances where MRs work and where they don't. But issue-driven science becomes normative science when a researcher holds a bias towards a certain result's outcome which may manifest itself at any stage of research such as, the study design, data collection and analysis, interpretation and publication (Fanelli 2012). Researcher bias may arise because of self-interest (funding, to get published in high impact journals) or for normative reasons (a preferred policy drives researchers to undertake studies that are likely to support that preference (Tomkins and Kotiaho 2004), and to show that policy in a favourable light). In these circumstances, the blurring of the boundary between science and policy advocacy becomes problematic.

4.2 Pro-MR Bias

A short questionnaire was sent to 200 leading scientists who have studied the ecological effects of MRs, to test for such biases. Ninety two questionnaires were completed, providing anecdotal evidence from which the following conclusions were tentatively drawn. A number of respondents thought that a publication bias for positive MR effects existed in the literature because of two reasons: (1) it takes more time and effort to get a non-significant result published; and (2) in conceiving a research project, scientists may frame their study in a way that maximises the likelihood of significant results. Whilst a few authors believed they had experienced ideological bias for MRs in unfavourable reviews of their work, it is unlikely that this would have had an effect on the balance of evidence produced in the literature since affected authors indicated that they managed to get their work published elsewhere. But the question of whether pro-MR views have led researchers to cross the boundary between science and policy advocacy remains to be answered.

Table 4.1 Description of the categories that were used to code each scientist's response

Reason for rejection	Description
Methods/quality	Unsound methodology, poor write-up
Interest	Insufficient interest to the journal, because results were too specific (not generalisable) or not novel enough
Ideological bias	Paper rejected because reviewers/editor thought findings would affect the MR cause
Personal bias	Paper rejected because of reviewer's/editor's personal agenda against the author or competing research programme

4.2.1 Methods

The search string 'Marine AND ("reserve*" OR "protected area*" OR "park*" OR "sanctuary*" OR "no take zone*" OR "conservation zone*" OR "refugia" OR "closed area*")' was used to source all literature published on MPAs (1972–2010) from ISI's Web of Science (WoS). Records were imported into HistCite^{TM} and authors ranked according to their publication count. The 200 leading scientists in MPA research were identified through their number of publications. The questionnaire was initially piloted to ten scientists randomly chosen from this sample, and a few of the questions were reworded after some suggestions by respondents. The final questionnaire was sent out via e-mail to the remaining 190 scientists between April and June 2012. If a scientist had not responded within 1 month, a reminder e-mail was sent with the questionnaire reattached. The questionnaire comprised 11 questions that were designed to explore a scientist's experience with publishing ecological effects studies on MRs. Questions were deliberately broad and left open to interpretation (e.g. on the issues of what is meant by 'bias' or 'positive effect') so as not to lead the respondent.

The questionnaire's primary purpose was to determine if a scientist had experienced rejection of a paper on MPAs, and if so, what reasons did the scientist believe were behind the paper's rejection. In addition, questions were asked about whether scientists chose not to submit particular research findings, and if so, their reasons for doing so. From an examination of the reasons given by those scientists who had experienced a rejected paper for why their work had been turned down, we identified four categories: methods/quality, interest, ideological bias, and personal bias (see Table 4.1). These categories were not mutually exclusive, and sometimes scientists gave or hinted at two reasons for the rejection of their paper; in such cases both reasons were tagged.

Although scientists were chosen for this questionnaire survey because of their publication count, it was not possible to know whether the views of this sample of scientists were representative of the whole scientific community that has studied MRs, so we cannot quantify the extent of any 'bias' occurring in the MR scientific literature.

Fig. 4.1 Reasons given by author for their paper's rejection

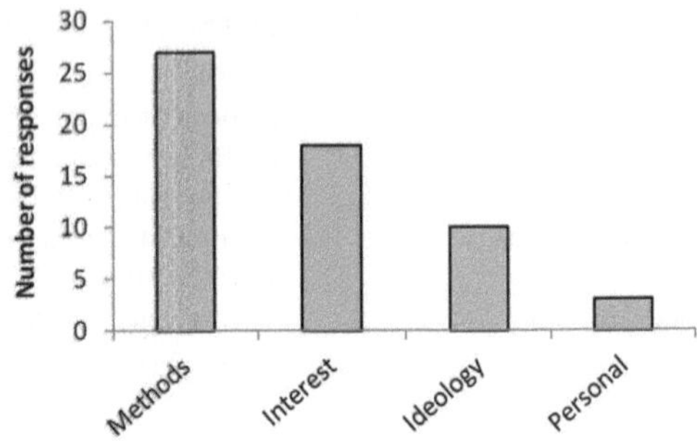

4.2.2 *Results and Discussion*

One hundred and four scientists out of 200 responded to the survey; the largest number (42 %) came from the USA, with the remainder from Spain (9 %), Australia (8 %), Canada (8 %), New Zealand (8 %), UK (8 %), and Italy (7 %). Twelve scientists did not complete the questionnaire but did express their viewpoints, so in total 92 fully completed questionnaires were collected. Of these, 50 scientists (54 %) said that they had experience of papers on MRs being rejected.[1]

Journal Editor and Reviewer Bias

Of those scientists who gave a reason for their paper being rejected, most felt that the reasons were methodological or lack of interest rather than bias (see Fig. 4.1). Nevertheless, 11 % believed that their paper had been rejected for ideological (i.e. normative) reasons.

1. Paper rejected for ideological reasons
   Whilst Fig. 4.2 should not be over-interpreted due to the very small sample size ($n=10$) and statistical non-significance of the ten scientists who thought their paper had been rejected because of ideological bias, seven suggested that this was due to the bias of the reviewer/editor against a negative/non-significant result, and three suggested that it was the result of reviewer's/editor's bias against a positive result (see Fig. 4.2). Both sets perceived biases—towards and against MRs—providing some impressionistic evidence for the view of some authors that the scientific community has become politicised with regards to the use of the MR as a management tool (Agardy et al. 2003).

Although the quotations pasted below are anecdotal, they are used as qualitative data to illustrate the fact that some scientists surveyed thought they had experienced ideological bias from peer-reviewers or editors. One scientist said:

[1] Quotations from respondents are italicised.

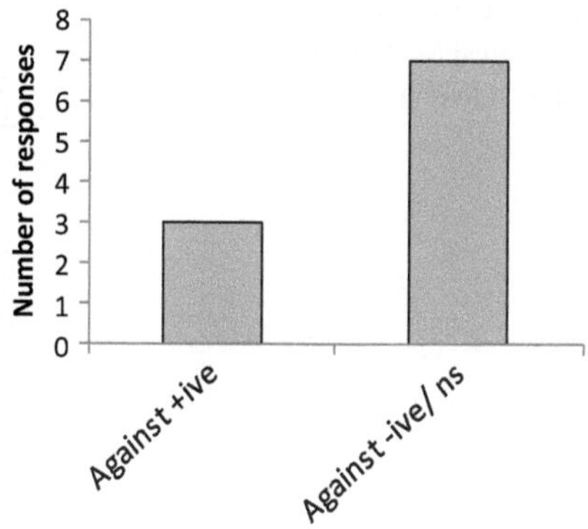

Fig. 4.2 Scientists' perceived political bias amongst editors for their paper being rejected

> *We were saying MPAs did not prevent declines in coral cover and associated declines in fish richness and abundance. One of the reviews seemed to circulate around the fact that we could not/should not be publishing bad news stories about MPAs.*

Another said:

> *Well, I do think that one reviewer had a publication bias against papers showing that MPAs don't work...s/he made a remark along the lines that s/he did not feel a paper showing that MPAs don't work should be published in a high-profile journal.*

- though the same scientist did suggest that there may also have been some legitimate methodological reasons why the paper was rejected, and that it was only one reviewer who displayed a pro-MPA bias:

> *To be fair, there were some issues that we addressed before re-submitting to the next journal and quite frankly, I don't think of this as a systemic bias, but rather one poor reviewer and that journal typically rejects anything that does not have a unanimous consensus among reviewers.*

Another scientist, whose theoretical work challenged some of the assumptions made about the beneficial impact of MRs on fisheries outside the protected area (the spillover effect), expressed his frustration at having his manuscript rejected by three different conservation/ecology journals:

> *For our theoretical paper on spillover these reviews completely missed the point of our manuscript and provided incorrect technical critique as a justification for why they were not considering our manuscript...I am sure the associated editors fully believe that their critiques are correct, but it seems they have read our theoretical paper with their own preconceptions at the front of their mind and no matter how explicit we are, we cannot break them down. I also get the feeling that because we are challenging beliefs about MPAs that may undermine some of the previous evidence about their benefits (i.e. spillover)...it is especially hard to communicate our point.*

Another respondent hinted that a bias against articles which find that MRs do not generate a spillover effect existed in conservation journals which were not interested in whether or not MRs benefitted fisheries:

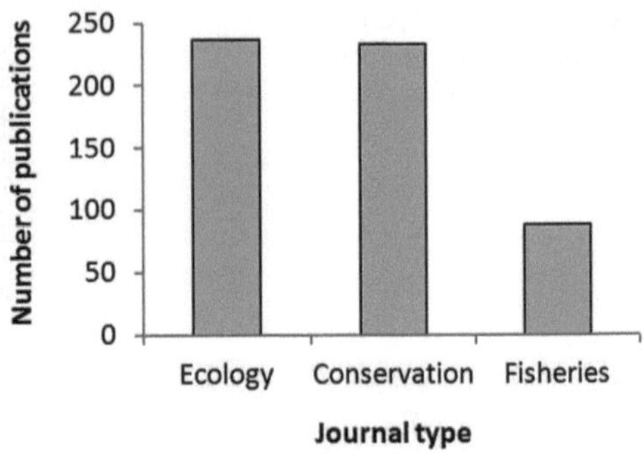

Fig. 4.3 The total number of ecological MPA studies by general journal type

MPAs have been first implemented with a conservation focus rather than a fisheries focus.[2] *Consequently, the MPA literature has had a strong conservation slant for some time. (see* Fig. 4.3 *and* Table 4.2*)*

One respondent took a benign view of the whole controversy, arguing from a pluralistic perspective that one bias balanced out another:

I don't think there is this bias. While each individual carries a bias, this is balanced by others with opposite bias. On average, there is no or very little bias in peer review, taken as a whole. That's why, if a paper is rejected in one case, another journal with different reviewers will likely respond differently, as long as the underlying science etc, is up to scratch.

This argument about balance receives some support from the fact that one respondent detected an anti-MR bias in one prestigious journal:

A very clear anti-MPA bias on the part of the editors (one of whom publishes anti-MPA papers).

2. Paper rejected because of insufficient interest
   The journal an author initially submits to is likely to have a relatively high impact factor and consequently high rates of rejection, with even methodologically sound papers being turned down if they were not of sufficient interest (see Table 4.3).

One scientist said:

For example, our best work, submitted to Ecology, Science and Nature, was rejected, not on the grounds of quality of the science; rather, they felt that MPA-related issues were not enough of popular interest to their readership.

Whether a particular finding is deemed enough of popular interest may depend on external political factors. For instance, the salience of an issue on the political agenda (in this case overfishing and MRs) may have more influence on whether an editor chooses to accept an article than does the quality or findings of the paper (Hilborn et al. 2004; Hilborn 2007a). The definition of sufficient popular interest may be used synonymously with 'importance'. One MPA study that did get published in

[2] Yet, the predominant focus of the empirical literature on MRs is their effects on fish (see Chap. 5).

Table 4.2 The top 20 journals where ecological studies of MPAs have been published. Each journal is labelled according to its general audience: *E* Ecology, *C* Conservation, and *F* Fisheries

Journal	Type	Publications	Impact factor (2011)
Marine Ecology Progress Series	E	92	2.7
Ecological Applications	C	52	5.1
Aquat Conserv: Marine Freshwater Ecosystems	C	44	1.9
Biological Conservation	C	41	4.1
ICES Journal of Marine Science	C	40	2
Can Journal of Fisheries and Aquatic Sciences	F	39	2.2
Bulletin of Marine Science	E	39	1.1
Conservation Biology	C	34	4.7
Fisheries Research	F	27	1.6
Environmental Conservation	C	22	1.9
Marine and Freshwater Research	E	18	1.6
Coral Reefs	E	17	3.9
Biologia Marina Mediterranea	E	13	1.4
Ecology Letters	E	13	17.6
Marine Biology	E	13	2.3
Fishery Bulletin	F	12	1.1
Journal of Exp Marine Biology and Ecology	E	12	1.9
Journal of Applied Ecology	E	10	5
Reviews in Fish Biology and Fisheries	F	10	2.5
African Journal of Marine Science	E	10	1

Science and received considerable press attention was that by Roberts et al. (2001). One of the co-authors of this paper initially said:

> *our work on the effect of the Merritt Island reserve in Florida on the catches of trophy fish was initially rejected by Science, however when it was combined with work by Callum Roberts on the effect of reserves on coral reef fishes, the combined work achieved the journal's threshold of importance.*

One of the key dimensions of importance is that a result is deemed to be generalisable beyond the local or regional level. For example, two respondents gave the following reasons why their paper was initially rejected:

> *referees and editors don't like local papers*

> *the result was too regional, and not important enough*

Another criterion of importance is a study that shows a 'statistically significant effect'[3]: one scientist (who is also a journal editor) was quite candid about rejecting MPA studies that showed no significant effects:

[3] Some studies have suggested that there is a bias in higher impact factor journals towards studies that show stronger effects sizes (Barto and Rillig 2012).

Table 4.3 Description of interest criteria (derived from questionnaire responses) that may affect the likelihood of a study being published in a top journal

Criteria	Description
External political reality	Salient political issues may have some bearing on whether an editor chooses to accept or reject a study
Generalisable	Result is likely to be applicable in a variety of contexts. Studies carried out at the local scale may only succeed in being published in a local journal
Statistically significant	Result shows that a positive effect is highly likely to be attributed to the effect of protection
Strong effect	The stronger the effect size the more likely the finding will be published in a top journal
Counter-intuitive	Result contradicts a previously held belief

> *As an editor of leading journals it is common to send back papers with no significant effect and ask the authors to send to a lower impact journal. This is just the hierarchy that exists in journals and is well known by those who handle papers. Nevertheless, a very well replicated and designed study that shows no effect on something that has been commonly stated as having an effect would also attract these same high impact journals.*

As the last sentence in this quotation suggests, a study that showed counter-intuitive results—such as indicating that an MR has non-significant or negative effects on a species that has been shown in other studies to increase—might attract publication in the top journals. On the other hand, if a belief is so deeply engrained within the scientific community, a counter-intuitive result may be virtually impossible to get published.

3. Paper rejected for methodological reasons
   There are several legitimate methodological reasons why a paper showing non-significant MR effects may be rejected. For instance a paper may not replicate treatment or may not control for factors such as time, age, poaching or recreational fishing activity—all of which will have a strong bearing on the study outcome. However, one author said that even here there may be a bias, in that reviewers may be much more critical of methodological flaws in papers that denigrate MRs:

> *I think in most cases you will not find clear cut cases of rejection just because papers have null or negative results for MPAs. Rather these factors raise the bar for acceptance and make reviewers more likely to attack other weaknesses in the paper (that always exist in any publication)…Often MPA papers that are less than flattering get knocked down a notch—rejected instead of revisions, revisions instead of accepted. Generally, journals advise publishing in a more 'specific' journal that always has a lower impact factor*

So MR studies that show strong positive effects may be less severely scrutinised methodologically in high impact journals (Hilborn et al. 2004), despite having flawed designs and poor data quality on which to draw robust inferences (Willis et al. 2003a). This causes a bias in the literature towards studies that show positive effects, which is problematic when meta-analyses[4] are conducted to make generalisations about the strength of MR effects:

> *One meta-analysis clearly demonstrated a strong bias in MNR-related publications towards only positive results, the severe failure of studies to employ a BACI approach, and the very selective focus of many studies towards focusing only on a species expected to change*

4. Paper rejected for personal reasons
   Two respondents indicated that they thought their paper had been rejected because of personal competitiveness from some of their peers, but not because of any pro- or anti-MR bias. One of them said:

> *alleged poor quality, e.g., the assholes did not like natural history and worse, we know who they were, and our paper was much better than anything that they have published! We finally got it published and it is an excellent paper, but positive MPA effects had nothing to do with the rejections. NO (not due to the result outcome), it was more personal and competitive.*

Another respondent suggested that an editor had a personal grudge against them, underpinned by normative differences:

> *I have been told I am not a true conservationist, and a friend of the enemy!*

Author Self-Censorship

Values may affect a scientist's everyday decision making about what tasks they choose to prioritise over others. Of significance to this study is whether some authors self-censor their results because they fear the results will have negative repercussions for a preferred policy. Of the 92 respondents, 76 said they had submitted or prioritised work on MRs that showed non-significant or negative effects, but 16 said they had not (see Fig. 4.4). The latter figure is likely to be an underestimate, because scientists may be unwilling to divulge that they self-censor; two respondents strongly abhorred the practice:

> *No! No self-respecting scientist would do anything like that.*

> *No, this would be clear bias by the scientist and be a violation of professional ethics.*

One respondent claimed to know of self-censorship by colleagues who were totally committed to the value of MRs:

> *I have never personally had a paper rejected that was critical of MPA policy or practice, but I have seen evidence of self-censorship by scientists who believe strongly in the value of MPAs that they refuse to say anything critical of their use or their success. My experience*

[4] It has been shown in other research fields in ecology that meta-analyses, in seeking to make generalisations from multiple studies of single scientific phenomenon, are often skewed by such publication bias (Murtaugh 2002). Therefore, the claims made by meta-analyses on the universality of MR effects must be treated with caution.

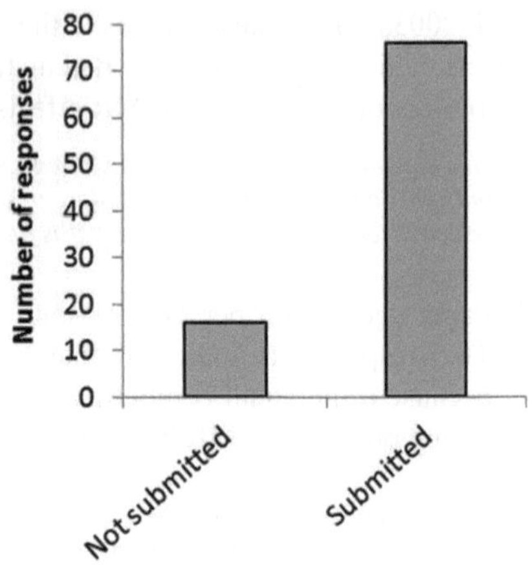

Fig. 4.4 Number of scientists who admitted that they did not submit or prioritise work showing non-significant or negative MR effects

of this attitude came in disagreements with co-authors on how to say things we wanted to say in a paper, and I am sure the attitudes that warned against criticism would also cause them to soft-pedal any negative data they had.

Of the 16 respondents who admitted self-censorship, two said that they did so because of pro-MR bias in the journals:

Likely rejection because of the results was definitely a factor in the decision. In some cases, it felt a bit like a Don Quixote-esque battle with wind mills, promising a long battle for publication and potential for exclusion from certain collaborations because of the perception of my not being 'onboard' when it comes to MPAs.

There may be more people working on illustrating/reinforcing their intuitions that MPAs are always beneficial than there are people working to illustrate any negative impacts... there is some kind of band wagon that many people appear to believe they must jump on if their work is to be published.

Another respondent defended the practice of self-censorship on grounds of priorities:

Yes, that is normal because it is not going to interest the top journals, but most studies show effects. One is trying to get into leading journals and then I am very busy, so putting effort into a paper that will not get into a leading journal and that takes so much time is not a high priority.

Summarising the findings of our investigation into the claims of a pro-MR bias in the peer-review process, we can say that although there is little hard evidence of such a bias, anecdotal data show that some scientists believe that pro-MR papers receive more favourable reception than do anti-MR papers in some journals. The question arises whether pro-MR scientists have allowed their commitment to the MR cause to make them cross the borderline between science and policy advocacy.

4.3 The Science/Policy Divide

The point of maintaining the so-called science/policy divide is to prevent science from losing its integrity or objectivity by being swayed by the political exigencies of the policy arena. This is the linear model of science, according to which, scientists should give their findings to policy makers impartially and without fear or favour, uninfluenced by any political agenda that policy makers may have and leave decision-making to the policy makers. Accordingly, scientists should not be tempted into the political arena as advocates of any particular policy choice because such temptation might lead to the scientisation of politics, whereby the scientists' advice would be based on their subjective, value-laden, political preferences, attempting to frame the policy debate in a way that precluded discussion of other values. Correspondingly, policy makers should accept the impartial advice given to them by scientists, and not try to influence scientists to bend their advice to suit the political agendas of the policy makers, which would amount to the politicisation of science.

An alternative interpretation of the proper relation between science and policy is the deliberative model, whereby scientists are regarded not as ivory tower elites standing up for objective truth above the political fray, but as stewards or custodians of the public good, which entails entering the policy debate as advocates for measures which their research findings demonstrate to be imperative in the public interest. On this view, the science/policy divide evaporates as scientists become advocates, and policy makers search for scientists who will supply them with evidence to support their preferred policy choices. In the remainder of this chapter we examine whether pro-MR scientists have kept to the straight and narrow of the linear model of the science/policy divide or have embraced the deliberative model, and if so, whether it matters.

4.3.1 The Linear Model

Evidence that pro-MR scientists kept to the linear model includes their epistemic community (EpC) activity, which sought to provide policy makers with scientific information about MRs, but stopped short of entering the policy arena directly (see the right-hand column in Table 4.4).

4.3.2 The Deliberative Model

The deliberative model entails policy advocacy by scientists, directly entering the policy area. Policy advocacy is the pursuit of influencing outcomes—including public-policy and resource allocation decisions within political, economic, and social systems and institutions—that directly affect people's lives (Cohen et al. 2001) (see the left-hand column in Table 4.4). Many would argue that it is desirable for

Table 4.4 A continuum of policy advocacy with examples of actions that conservation biologists might take in conducting and reporting research. Actions on the left represent policy advocacy; those on the right do not; and those in the centre may or may not. (Taken from Scott et al. 2007)

Policy advocacy?		
Yes	Maybe	No
Stipulating preferred policy decisions	Using language and words in ways that can be interpreted differently by different groups or stakeholders	Conducting research on policy-relevant issues
Supporting a class of policies based on only general beliefs or values	Failing to acknowledge the full range of potential consequences of scientific uncertainty on interpretation of research	Publishing results in scientific journals
Conducting normative science	Sharing research results with one or a limited range of special-interest groups	Publishing results in non-technical outlets
Lobbying for specific policies or management outcomes	Providing advice to one stake-holder about a controversial issue	Bringing relevant science to the attention of managers and policy makers
Framing research questions or choosing study areas such that the outcome will support preferred policies		Providing results of research to all stakeholders and the public
		Supporting the use of the best available science in decision making
		Testifying before congressional committees
		Giving interviews to the press about research results
		Discussing conservation science on radio or television shows

scientists to become more engaged with the policy process (Steel et al. 2004; Gray and Campbell 2008), and there is a widely-held belief within the scientific community that science compels action (Davis 1999). In UK universities, the Research Excellence Framework (REF) requires academics to show the wider influence of their research on policy, thus pressurising them to become more proactive in the dissemination of their research findings to their peers and also to members of the wider policy community. This might lead to policy advocacy.

Much of the underlying basis for scientists' advocacy of MRs has been a normative commitment to protect biodiversity. Lackey (2001) claimed that, for example, disciplines like conservation biology and restoration ecology exemplify normative science by adopting value-laden precepts implying that biodiversity is intrinsically good, but he criticised such normative science: 'For the political process of adjudicating conflicts over value and preferences, science offers no moral or ethical guidance' (Lackey 2001, p. 444) Similarly, Boyd (2013) pointed out that in some policy areas, 'based upon the same evidence, it is possible to pursue quite different policy options. Unfortunately, some scientists have been drawn into the public debate about which policy option is correct. If scientists start to say that one or other option is right or wrong then they are beginning to take the position of politicians and they devalue the scientific evidence they claim to present'. Pielke (2007, p. 7) explained why normative science is an attractive option for scientists: it 'is politically desirable because it allows for a simultaneous claim of being above the fray, invoking the historical authority of science, while working to restrict the scope of choice. The Stealth Issue Advocate seeks to "swim without getting wet"'. There is some evidence that such stealth issue advocacy is pervasive in the ecological and natural resource scientific literature (Scott et al. 2007), but such normative science can lead to the politicisation of the scientific community whereby research is conducted and interpreted in order to influence policies (Agardy et al. 2003), which could undermine the credibility of scientists.

4.3.3 Is Scientific Advocacy of MRs Acceptable?

Christie (2011, p. 179) claimed that scientific advocacy by marine scientists is a recent development: 'The marine scientist as an advocate of social policy is relatively new and simultaneously applauded and vilified'. One answer to the question of whether or not scientific advocacy is acceptable is that if a scientist's advocacy for MRs is based on compelling evidence, his/her policy advocacy is legitimate. However, who is to judge whether the evidence is compelling? In the field of MR research, evidence is rarely compelling but nearly always contested. Perhaps the evidence is deemed to be compelling when there is a scientific consensus in favour of MRs? Policy makers are more likely to accept scientific advice if there is consensus among scientists. But there will always be dissenting voices and alternative ideas—indeed, scientific progress depends on disagreement (Sarewitz 2011). Nevertheless, although complete scientific consensus is rarely achievable, there may be sufficient consensus for legitimate advocacy of policy action. Attempts have been made to demonstrate such a consensus on MRs: indeed, consensus statements on MRs have been developed to create the political will necessary for MR designation. Whilst it is true that signing a consensus statement on MRs does not necessarily mean that the signatory is an advocate, if the statement strongly endorses a pro-MR policy commitment, it is difficult to avoid concluding that signatories are advocates. For example, the European Scientists' Consensus Statement on Marine Reserves organised by Callum Roberts in 2007 and signed by 275 scientists in Europe with masters or PhD qualifications constituted an unequivocal endorsement of the MR case: 'We, the undersigned, believe that Fully Protected Marine Reserves are essential for conservation…[and] may benefit fisheries by the "spillover" of animals from inside the reserves and from the export of eggs and larvae to adjacent marine areas…In order to assure sufficient protection across the whole range of marine ecosystems it is necessary to establish a representative, replicated, networked and

Table 4.5 Conclusions from the Scientific Consensus Statement on MRs. (NCEAS 2001)

Ecological effects within reserve boundaries
1) Reserves result in long-lasting and often rapid increases in the abundance, diversity, and productivity of marine organisms
2) These changes are due to decreased mortality, decreased habitat destruction and to indirect ecosystem effects
3) Reserves reduce the probability of extinction for marine species resident within them
4) Increased reserve size results in increased benefits, but even small reserves have positive effects
5) Full protection (which usually requires adequate enforcement and public involvement) is critical to achieve this full range of benefits. Marine protected areas do not provide the same benefits as marine reserves
Ecological effects outside reserve boundaries
6) In the few studies that have examined spillover effects, the size and abundance of exploited species increase in areas adjacent to the reserve
7) There is increasing evidence that reserves replenish populations regionally via larval export
Ecological effects of reserve networks
8) There is increasing evidence that a network of reserves buffers against the vagaries of environmental variability and provides significantly greater protection for marine communities than does a single reserve
9) An effective network needs to span large geographic distances and encompass a substantial area to protect against catastrophes and provide a stable platform for the long-term persistence of marine communities

sustainable system of Fully Protected Marine Reserves' (Roberts 2007a). Similarly, the Scientific Consensus Statement in NCEAS (2001, p. 3) signed by 161 leading marine scientists and experts on marine reserves claimed that 'existing scientific information justifies the immediate application of fully protected marine reserves as a central management tool'. Table 4.5 sets out the conclusion from the NCEAS consensus statement on MRs.

However, most of the signatories to this 2001 consensus statement were marine ecologists (Hilborn 2007c), which raises questions about the strength of the evidence base for the fisheries effects of MRs (particularly points 1 and 6 in Table 4.5; Agardy et al. 2003; Willis et al. 2003a). More worryingly, such consensus/majority statements may have used science to gloss over what is arguably a heavily value-laden debate, and if so, this would be a form of stealth issue advocacy and an illustration of 'politicizing science', which occurs whenever someone invokes science as a justification for selecting one course of action over others (Pielke 2007). Critics argue that sometimes within positivist science lie unacknowledged normative assumptions which result in scientific advice being developed, presented, or interpreted with a tacit preference for particular policy choices (Lackey 2007; Jones 2002).

Some defenders of scientific advocacy for MRs are prepared to justify it even if the consensus is based on normative assumptions rather than scientific objectivity. For example, Jones (2014, p. 98) defended MR advocacy, acknowledging that papers that are written to support MRs are normative in that biologists who examine the decline in health of marine ecosystems 'cannot base their arguments solely on objective, value-free, positivist science': indeed, Jones confessed himself to be an

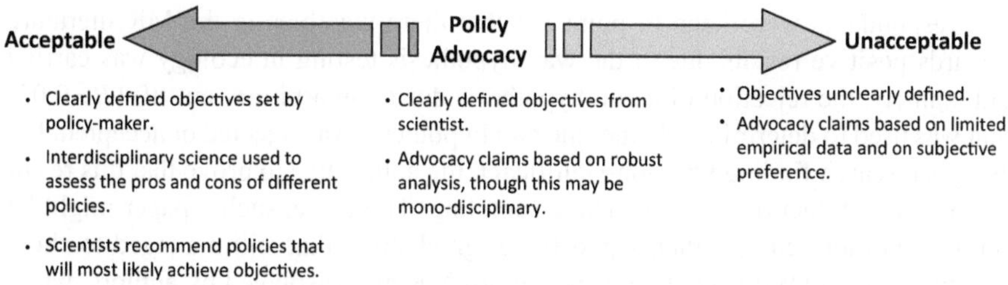

Fig. 4.5 Circumstances when scientific advocacy is acceptable or unacceptable

MR advocate. However, critics of scientific advocacy argue that where such advocacy reflects normative assumptions, it has crossed a red line between science and policy by using science to support subjective preferences. Pomeroy et al. (2005) criticized MR advocacy that exaggerated its case by stepping beyond the science.

Another view is that scientific advocacy may be acceptable if scientists work closely with policy makers to identify possible solutions to pre-specified policy problems (see Fig. 4.5). Such collaborations are a common experience for scientists, indicating that 'good' policies do not usually arise from science in isolation (as the linear model suggests) but through a process of multi-way communication between society, policy makers, and scientists (Nowotny et al. 2001, 2003) (as the deliberative model suggests). However, critics of the deliberative model argue that during such collaboration, policy makers may put pressure on scientists to play down scientific uncertainty. Also, studies have shown that there may be pressure exerted on scientists from certain stakeholder groups (such as environmentalists) to become advocates (Gray and Campbell 2008). One way to avoid these dangers might be to require that scientists, when recommending policies to decision makers, should always provide a range of options so they are not reducing the scope of choice (Pielke 2007).

4.4 Conclusion

This chapter had two parts: first, an investigation into the allegation that there was a pro-MR bias in the peer-review system; and second, that this bias spilled over into scientific advocacy for MRs that breached the barrier between science and policy. On the issue of pro-MR bias, there were two components: bias displayed by editors/peer reviewers, and bias displayed by authors. On pro-MR bias displayed by editors/peer reviewers, only a small minority of respondents ($n = 10$) suggested that their paper had been rejected because of an editor/peer reviewer's belief that a study showing non-significant or negative findings would damage the MR cause (defined here as an ideological bias). Moreover, these respondents managed to get their work published elsewhere which would suggest that any ideological bias by the editors/reviewers would have had little effect on the literature overall. Nevertheless, several respondents were keen to point out that there was bias in the MR literature towards positive results due to the way hypothesis testing in ecology was carried out: namely, the rejection of a null hypothesis due to an arbitrary cut-off ($P > 0.05$) that was used to determine whether the null hypothesis was rejected or accepted. If a study's research finding was non-significant, the author had to prove that this result was not an artefact due to insufficient sampling. Therefore, such a paper might be scrutinised more closely than a poorly-designed study that still managed to show a positive effect (Willis et al. 2003a). On pro-MR bias displayed by authors, while no author openly admitted to self-censorship, one scientist believed that some of his peers had self-censored results because of their belief that it could damage the cause of MRs. Moreover, a number of scientists admitted to not submitting non-significant results due to the perception that such a result would not be of interest to the high impact journals to which they prioritised their time to achieve publication. Therefore many scientists deemed it not to be worth the effort in trying to get non-significant results published.

The first part of this chapter has shown therefore, that there is anecdotal evidence that a pro-MR belief system exists amongst some members of the scientific community who studied MRs, and although it is likely to have led to relatively little significant bias within the MR literature for studies documenting positive effects, this does not rule out the possibility that such a bias might occur at a sub-conscious level, perhaps akin to Kuhn's idea of a paradigm, a set of widely-held assumptions in the scientific community that made it hard for any counter-intuitive result to be conceived by researchers, let alone taken seriously by editors (Kuhn 1970; Koricheva 2003). Nor does it rule out the possibility of an institutional culture of scientific incentives which pressure researchers to maximise opportunities for citations by a 'file-drawer' strategy whereby non-significant results remain unpublished.

In the second part of the chapter, the issue was raised of whether pro-MR scientists have crossed the line between science and policy advocacy, and if so, whether it mattered. Here we found that while some scientists held fast to the linear model of the EpC whereby scientific findings were conveyed impartially to policy makers, others adopted the deliberative model of engaging closely with policy makers in a joint process of decision-making based on data that was incomplete, and on normative assumptions that were contestable. In assessing whether this was acceptable, we noted that where there was a consensus of scientific opinion, such scientific advocacy provoked less controversy, but that to guard against group-think, there should be a requirement that advocacy scientists provide a range of scientific recommendations for policy.

The next chapter looks at whether the claimed fisheries benefits of MRs are justified.

Chapter 5
Critique of the Scientific Evidence for Fisheries Benefits of MRs

5.1 Introduction

Three sets of generalisations about the benefits of MRs are heard from nature protectionists (NPs). First, NPs claim that MRs provide biodiversity benefits. For example, Grorud-Colvert et al. (2011, p. 293) asserted that 'Growing scientific information has shown consistent increases in species density, biomass, size, and diversity in response to full protection inside reserves of varying sizes and ages located in diverse regions' (see also Lester and Halpern 2008; Goni et al. 2011). Second, NPs claim that MRs generate fisheries benefits. For example, Geoffrey Lean (2009), the environment correspondent of *The Independent*, stated that 'Establishing 'no-take zones'...has been strikingly successful around the world; marine life has rapidly recovered and spread to surrounding areas, greatly increasing fish catches'. Third, NPs claim that MRs provide socio-economic benefits. For example, Fletcher et al. (2014, p. 264) held that 'There is evidence that...MPA networks have successfully been used for both fisheries management and to increase social and environmental welfare'.

But how far does the scientific evidence substantiate these three sets of claims? We accept that there is strong evidence that MRs increase abundances of some vulnerable species within their borders, which might be interpreted as enhancement of biodiversity; but we argue that the evidence that MRs substantially enhance fishery stocks is weak; on socio-economic benefits, the evidence is virtually non-existent. In this chapter, we focus mainly on the evidence of fisheries benefits, analysing the literature to assess the strength of the claim that MRs benefit fisheries. Building on previous studies of MRs (Willis et al. 2003a; Sale et al. 2005; Edgar 2011), we provide a quantitative overview of the ecological literature from 1970–2010, showing which ecosystems, marine reserves, habitat types and species have been most studied, and what are their implications for the fisheries management objectives of MRs.

A. Caveen et al., *The Controversy over Marine Protected Areas*,
SpringerBriefs in Environmental Science, DOI 10.1007/978-3-319-10957-2_5

5.2 Methods

5.2.1 Data Collection

The following word string—Marine *AND ('marine reserve*' OR 'marine protected area*' OR 'marine park*' OR 'marine sanctuar*' OR 'no take zone*' OR 'special area* of conservation' OR 'conservation zone*' OR 'specially protected area*' OR 'refugia' OR 'box' OR 'closed area*')*—was used to source all records published on MPAs between 1970 and 2010 from ISI's Web of Science (WoS) online interface. Eight hundred and thirteen ecological studies were identified after manually checking through abstracts to confirm that the MPA term was the main focus of the study; socio-economic and governance literature was excluded as beyond the scope of this analysis.

5.2.2 Literature Classification

The literature was classified into three categories: (1) empirical ($n=448$); (2) theory ($n=193$); and (3) reviews/notes ($n=172$) (though the last category was not analysed) (Willis et al. 2003a).

1. In the empirical literature, each study was categorised as 'MPA effect', 'MPA design', or 'methodological'. This was done to distinguish field studies that had measured an effect of protection ($n=310$) from field studies that had collected data on species distributions (Vanderklift et al. 1998; Curley et al. 2002) and species movements ($n=123$) (Holland et al. 1996; Meyer et al. 2000; Willis et al. 2001; Chateau and Wantiez 2009) aimed to inform MPA design, and from methodological studies that focused on monitoring MPAs ($n=15$) (Mouillot et al. 1999; Rudershausen et al. 2010). Only the 'MPA effect' studies ($n=310$) are analysed in detail in the results section. All MPA effect studies were categorised by marine ecoregion or marine province using the Marine Ecoregions of the World (MEOW) biogeographic framework (Spalding et al. 2007) and these were defined as 'tropical' where coral reefs were present; 'warm temperate' when average winter sea surface temperatures (SSTs) exceeded 10 °C; and 'cold temperate' where average winter SSTs were <10 °C. Studies that were undertaken in the Arctic and Southern Ocean realms were classified as 'polar'. Habitat was categorised in terms of the dominant substratum as hard or soft[1]. Most studies concentrated on one or other of these two gross habitat types, but when both hard and soft habitats were sampled, the study was counted twice. Occasionally studies did not explicitly state habitat type, and here habitat type was inferred from study species (e.g. lobsters associated with reef) and/or areas (e.g. estuarine

[1] In reality the hard–soft dichotomy is an oversimplification. Here hard habitats were defined as reef, and soft habitats as everything else (though this would also include a diversity of bottom types from mud and sand through to gravel and cobbles).

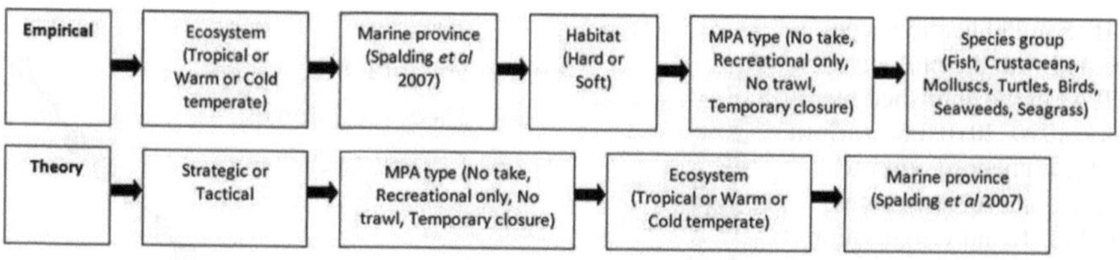

Fig. 5.1 Classification scheme for the empirical and theoretical MPA biological literature

and offshore areas were considered to be soft). The type of MPA studied was recorded: MRs were distinguished from MPAs that placed restrictions on only certain users (e.g. no trawl, recreational fishermen only) and temporary closures. The focus species of the study were also recorded (e.g. coral reef fish, soft-bottom fish community, pelagic). The 'quality' of MPA effect studies was also assessed through recording whether a study measured the effect of protection over a period < 1 year ('snap-shot' studies), > 2 years ('time series'), or had gathered information before and after an MPA had been implemented ('before–after'). Also recorded were studies that had measured habitat and used this as a co-variable in their analysis. The number of empirical 'MPA effect' studies was recorded for each of the 62 marine provinces (Spalding et al. 2007). This information was incorporated into ArcGIS 9.3 to show visually from which marine provinces most of the empirical evidence has come, and in which regions evidence is currently lacking.

2. The theory literature was classified as either 'strategic' or 'tactical' (Gerber et al. 2003). Strategic models have been developed to answer broad, overarching questions, such as what proportion of a given area should be placed in the reserve system? how many reserves are needed? and which types of data are most critical to obtain? Tactical models are more complex, containing details about specific situations, and used to inform local decisions on how MPAs can be designed to meet specific objectives (Gerber et al. 2003; see Fig. 5.1). The study sought to answer the following questions: (a) what is the proportion of strategic to tactical models? (b) for what types of ecosystem/species are strategic models most well developed? and (c) in which localities have tactical models been best developed?

5.3 Results

5.3.1 Empirical Studies

Number of Empirical Studies

As literature on MPAs has continued to expand exponentially, the proportion of empirical studies has increased relative to both theoretical and review/note type literature (see Fig. 5.2). This was observed for all empirical field studies that have been

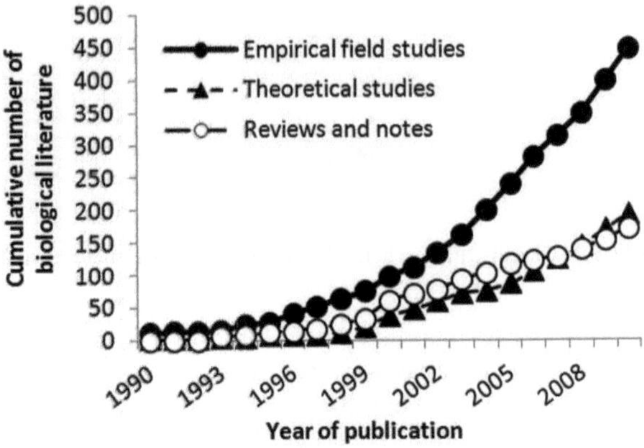

Fig. 5.2 Publications concerned with the biology of MPAs in the published literature, 1990–2010: comparison of the number of empirical, theory, and reviews/notes studies. To aid visualisation, papers 1977–1989 ($n=8$) were categorised as 1990

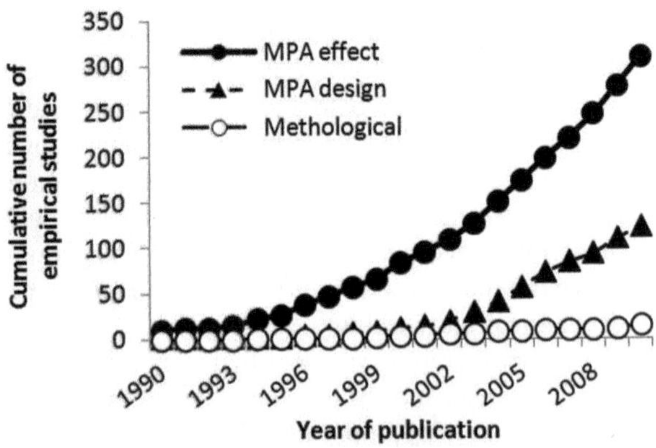

Fig. 5.3 Empirical studies broken down by type: whether they investigated the effect of protection (MPA effect); gained evidence on the movement/distribution of species/habitats to inform MPA design (MPA design); or investigated a methodological problem

undertaken in MPAs ($n=448$) and for those empirical studies that have examined the effect of protection only ($n=310$) (see Fig. 5.3). Recently, however, the annual publication rate of theoretical papers has been increasing at a similar rate to that of empirical studies, due to a leap in the publication of modelling studies.

Type of MPA Studied

Eighty-seven percent of the empirical literature has focused on effects of MRs (see Fig. 5.4). The remaining 13 % of studies have mainly focused on effects of no-trawl areas predominantly established over soft-bottom habitats in temperate seas (e.g. Murawski et al. 2000; Jaworski et al. 2010), or MPAs that only allow recreational users (Denny and Babcock 2004; Shears et al. 2006). Twenty-five percent of the empirical literature has come from the 10 MRs shown in Fig. 5.5. As of 2010, around 170 MRs had been studied, 30 of which were located in the Mediterranean Sea.

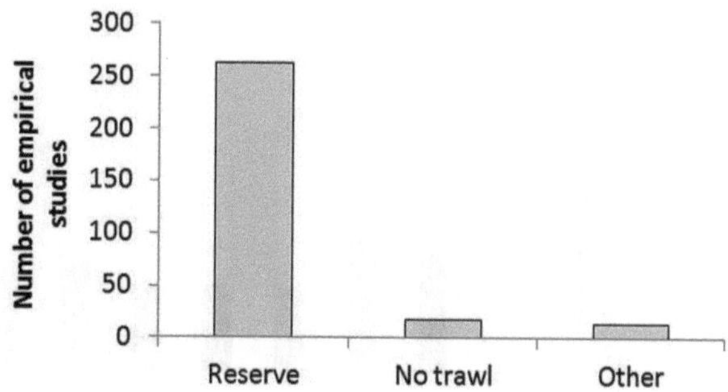

Fig. 5.4 Type of MPA studied: 'Reserve' defined as an area where no fishing occurred; 'No trawl' as an area where towed ground gear was prohibited; and 'Other' defined as areas that only allowed recreational users

Ecosystems and Habitats Studied

When the empirical literature is broken down by ecosystem type, more studies have been undertaken in the tropics ($n=119$) and warm temperate ecosystems ($n=116$), and more than half of this research effort has been undertaken in 25 MRs in the Mediterranean Sea. The publication rate of cold temperate ecosystems research lags well behind that of tropical and warm temperate ecosystems (Fig. 5.6). Reef-type habitats have been most studied (see Fig. 5.7a, $n=228$) with only 16 % of studies being carried out over soft habitats (see Fig. 5.7b, $n=43$ including no-trawl areas). There has been roughly the same research effort applied over soft habitats in warm temperate and cold temperate ecosystems, though this is mainly due to the study of the effects of large-scale groundfish closures that are predominantly located over soft ground (see Fig. 5.7b).

Main Groups of Species Studied

Reflecting the type of habitat surveyed, the main group of species studied were coral and rocky reef fish communities (see Fig. 5.8), which comprised 45 % of the focus species of all empirical studies. Reef crustaceans ($n=22$), coral reef fish predators ($n=22$) and molluscs ($n=19$) have been the subject of 20 % of empirical studies, while temperate soft-bottom fish communities have been the subject of only 5 % of empirical studies. No MPA effect studies were found for charismatic marine megafauna, although a few studies relevant to the design of MPAs did have empirical data showing the distribution and movement of turtles (de Segura et al. 2003), birds (Louzao et al. 2006; Terauds et al. 2006), and cetaceans (Canadas et al. 2005; Slooten et al. 2006) in MPAs and their surrounds.

Empirical Research Effort per Marine Province

Four marine provinces stand out in terms of the number of empirical studies that have been published on MPAs (see Fig. 5.9): the Mediterranean Sea (n=70); the

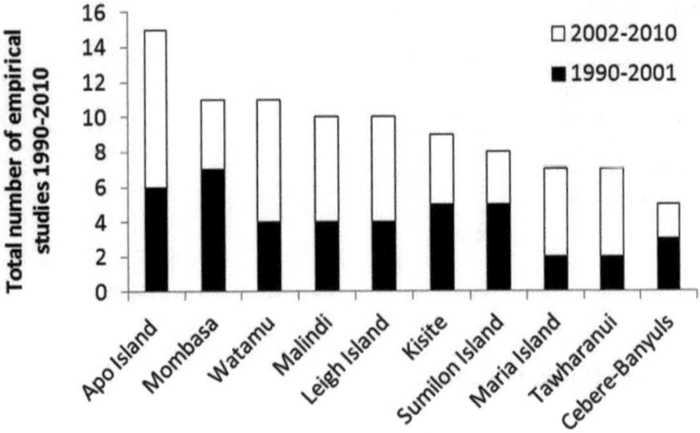

Fig. 5.5 Top 10 MPAs studied 1990–2010. Note that all are MRs

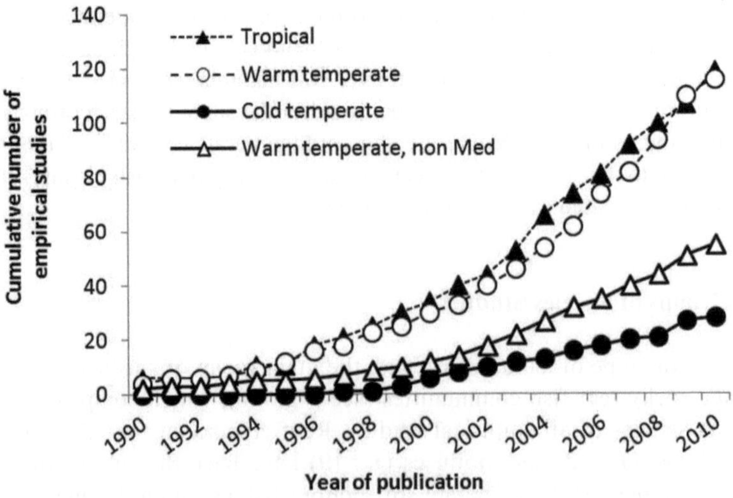

Fig. 5.6 Number of empirical field studies undertaken in MRs only by ecosystem type. The subset of warm-temperate studies conducted outside the western Mediterranean Sea is plotted separately

Tropical Northwest Atlantic ($n=33$); the Western Indian Ocean ($n=24$); and the Western Coral Triangle ($n=23$). Cold temperate marine provinces that have been moderately studied (between 6 and 11 publications) include Northern European Seas, the cold temperate Northwest Atlantic, and the cold temperate Northeast Pacific. Around half of the marine provinces have not been studied at all, and there are notable information gaps for Asia, Western Africa and the Southern Ocean.

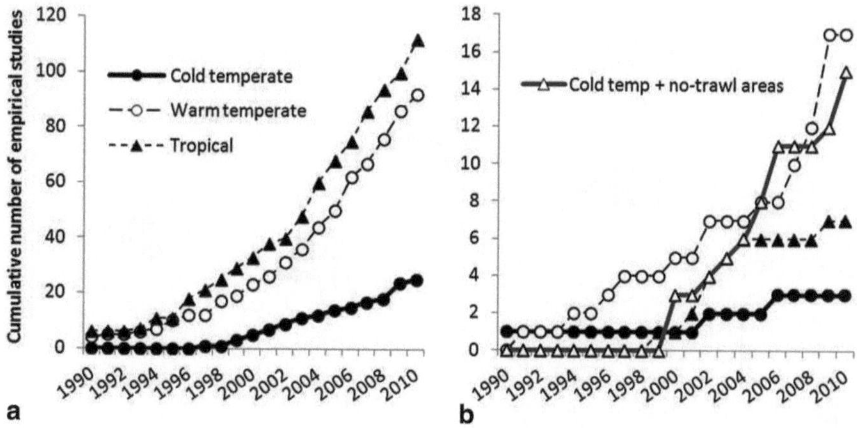

Fig. 5.7 Number of empirical field studies that have measured the effect of an MR over (**a**) hard (reef) habitats, and (**b**) soft habitats (with studies that have examined cold temperate no-trawl areas also shown)

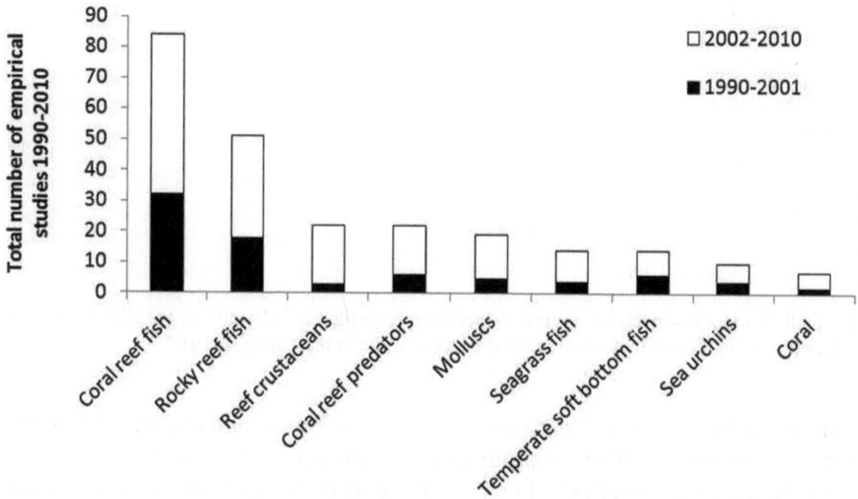

Fig. 5.8 Main focus organism(s) of MPA effect studies

'Quality' of Empirical Field Studies

Fifty four percent of empirical MR studies only presented a snap shot, having measured the effect of protection at a single point (i.e. a season) in time. Of these snap-shot studies, 25 % used only one fished control area to attribute a difference between sites to an effect of protection, and less than half of these explicitly tried to take into account effects of habitat in their survey design. Forty six percent of

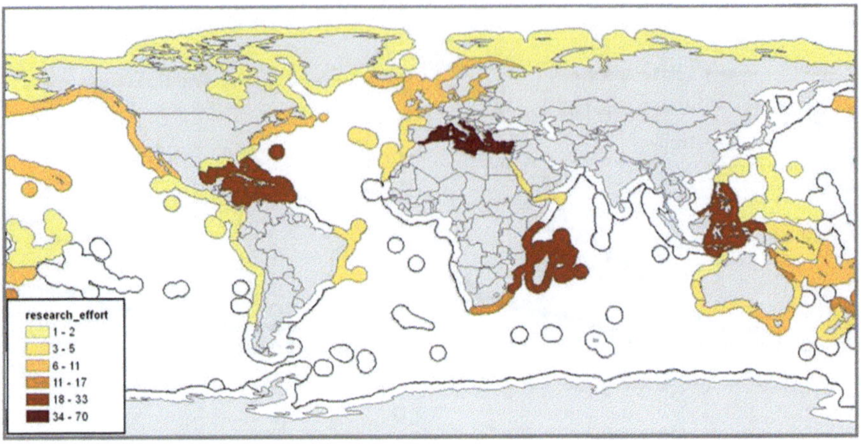

Fig. 5.9 Research effort (defined as number of empirical studies) per marine province: marine provinces with no colour have no MPA effect studies

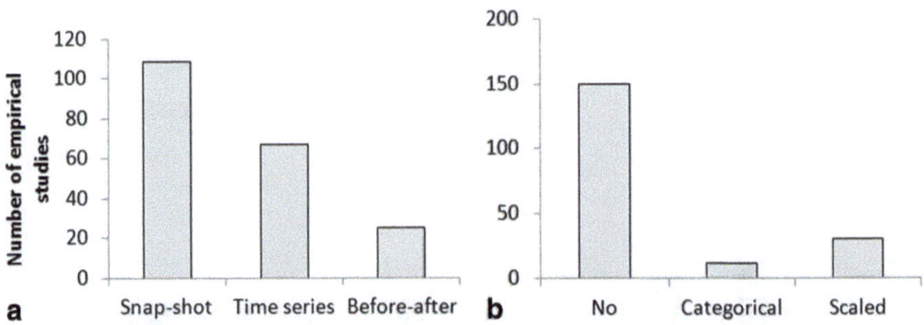

Fig. 5.10 Temporal aspects of empirical literature investigating MR effects (**a**) and number of studies that quantified habitat and used this as a covariate in their analysis (**b**)

studies took inter-annual variation into account in their design (see Fig. 5.10), but only 12 % measured effects of protection over time periods > 10 years.

Twelve percent of studies had before–after data (BACI); the design and focus of these studies are summarised in Table 5.1. The majority of before-after studies had measured effects of protection on abundance and biomass of coral reef fish species, though only one before-after study measured the effect of spillover (Francini-Filho and Moura 2008). Habitat was categorised (n=11) or scaled (n=30) by only 21 % of empirical studies.

Table 5.1 Summary of before-after studies of marine reserves, and number of temporal replicates per study (minimum two)

Reference(s)	Location	No. of reserves	No. of temporal replicates	Study focus
Russ and Alcala (1989, 1998, 2003, 2004)	Sumilon and Apo Islands, Philippines	2	2, 5, 13	Coral reef fish species richness, density and biomass
Bennett and Attwood (1991)	De Hoop, South Africa	1	3	CPUE [catch-per-unit-effort] of rocky reef fish assemblage
Wantiez et al. (1997)	New Caledonia	5	2	Coral reef fish species richness, density and biomass
Galal et al. (2002)	Nabq, South Sinai, Egypt	5	3	Density and size structure of commercially targeted grouper, emperor and snapper
Nardi et al. (2004)	Houtman Abrolhos Islands, Western Australia	4	6	Density of coral trout and wrasse
Claudet et al. (2006)	Couronne, France	1	3	Rocky reef fish species diversity and abundance
Hawkins et al. (2006)	St Lucia, Caribbean	4	7	Commercial coral reef fish species biomass
Lincoln-Smith et al. (2006)	Solomon Islands	1	6	Abundance and size of commercial coral reef invertebrate
Francini-Filho and Moura (2008)	Eastern Brazil	1	5	Biomass, size, and spill-over of coral reef fish

Table 5.2 summarises the factors that determine the strength of an MPA effect and which should be taken into account in the design of a monitoring programme.

5.3.2 Theoretical Studies

In the theoretical (i.e. modelling) studies, there was a greater abundance of strategic ($n=130$) than tactical models ($n=56$). The publication of strategic models increased rapidly during 1999–2000, and it was not until 2008–2009 that the publication trajectory of tactical models started to match that of strategic models—roughly a 9-year lag (see Fig. 5.11). Seventy percent of the strategic models were not calibrated to specific species: for those that were, 13 % derived their parameters

Table 5.2 Factors determining the strength of an MPA effect

Factor	Comments
Habitat heterogeneity	MPAs may be located over habitats that are comparatively resource rich (Hilborn 2002) or resource poor (Edgar et al. 2009). An ecological effect could be attributed to protection when it is instead due to habitat differences. To counter this, studies should use BACI designs (Underwood 1993) to increase the strength of their inference
Biological life histories	Individual species life-history traits strongly affect how they will respond to protection. Species may grow slowly so that any significant change will not be detected for years (Barrett et al. 2007). Alternatively, fish species may be highly mobile, so that an MPA has little or no protection effect (Shipp 2003). Survey designs and the amount of sampling effort needed to detect an effect will need to take species movement into account (Rotherham et al. 2007)
Environmental change	Regional environmental change and its effect on fish growth and recruitment may confound the interpretation of the effect of a closed area (Holland 2000). Such a problem could only be overcome through long-term monitoring
Past management history	The extent to which a fish stock has been exploited and habitat modified by fishing will influence the size of an ecological effect detected in an MPA relative to control locations and baseline at $t=0$
Current management history	Displaced fishing effort outside the MPA may lead to a greater intensity of fishing in its surrounds and lead the researcher to detect a greater effect of protection due to the deterioration of fish stocks and habitat outside (Hilborn 2002). Illegal fishing may also reduce the size of ecological effects (Bloomfield et al. 2012)

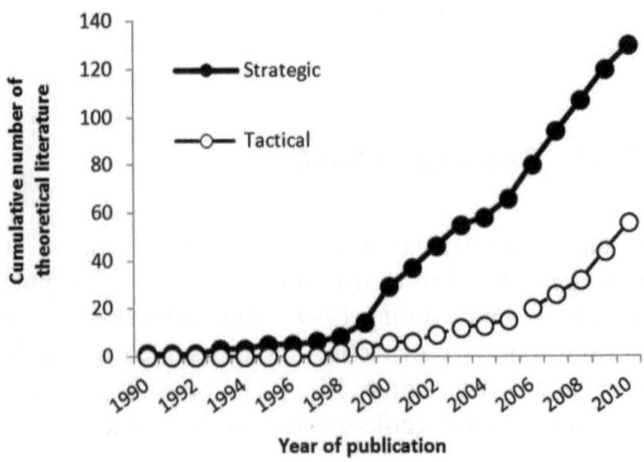

Fig. 5.11 Number of theoretical studies by model type

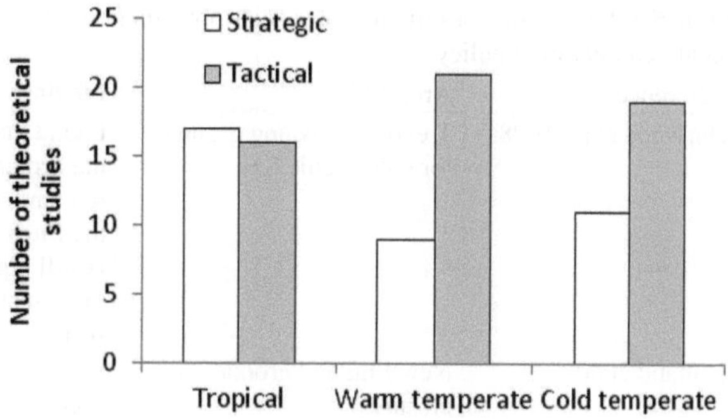

Fig. 5.12 Total numbers of theoretical studies by ecosystem. Note that 70 % of strategic models were completely abstract, whereas by their very nature all tactical models were developed for real-world problems

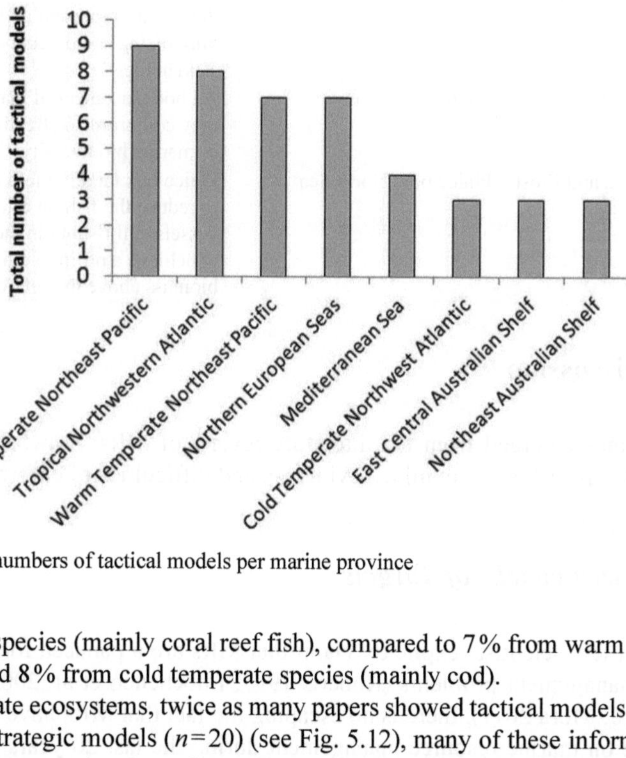

Fig. 5.13 Total numbers of tactical models per marine province

from tropical species (mainly coral reef fish), compared to 7 % from warm temperate species, and 8 % from cold temperate species (mainly cod).

For temperate ecosystems, twice as many papers showed tactical models (n=40) than showed strategic models (n=20) (see Fig. 5.12), many of these informing the design of MPA networks along the northeast Pacific coast of the USA (Ban 2009; Ban and Vincent 2009; Airame et al. 2003), and predicting the effects of groundfish closures (Horwood et al. 1998; Holland 2000; see Fig. 5.13 and Table 5.3).

Table 5.3 Some examples of tactical models potentially used to inform local fisheries and biodiversity conservation policy

Reference	Area/MPA	Findings
Horwood et al. 1998	Trevose spawning grounds for sole, Celtic Sea	Found that the closure of a sole spawning ground may be ineffective if sole remain free to be caught elsewhere, and the catch of sole outside the closed area is still high. Closed areas will be useful in protecting aggregations of juvenile fish
Holland 2000	New England groundfish closures	Impacts of closures will vary across species, sometimes increasing yields for some and decreasing yields for others
Airame et al. 2003	California Channel Islands	Identified reserve network scenarios that would represent all habitats whilst minimising socio-economic costs to stakeholders
Stewart et al. 2003	South Australia	Ad hoc placement of marine reserves may compromise effective conservation of marine biodiversity
Kjaersgaard and Frost 2008	Plaice box, North Sea	Plaice box largely ineffective; need to reduce the fishing effort of smaller vessels still fishing in the closed area to achieve a profitable fishery with a biomass above the reference point Bpa

5.4 Discussion

Four themes emerged from this literature review of MRs: drawbacks of targets; skewed focus of literature; mixed evidence; and difficulties of enforcement.

5.4.1 Drawbacks of Targets

Although few scientists explicitly claim that MRs are a panacea for all marine resource management problems (Roberts 1997; Lubchenco et al. 2003; Pitcher and Lam 2010; Mora 2011), there is no escaping the fact that MRs have dominated the literature on marine resource management during the last 20 years (see Chap. 1), and they have been advocated more frequently than other new management strategies such as marine spatial planning (MSP) or ecosystem-based management (EBM) (Halpern et al. 2010). Yet only 1.6% of the ocean is designated as MPAs and only 0.2% as MRs (Wood et al. 2008), which is a long way from meeting current international targets. In 2006 the CBD called for 10% of the oceans to be designated as MPAs by 2010, but, according to Wood et al. (2008), at the (then) current rate of progress, this target would not be achieved until 2050. However, a more up-to-date assessment by Spalding et al. (2013, p. 231), drawing attention to 'The dramatic acceleration of MPA coverage in recent years', in particular the increase in MPA designations of vast areas of the ocean during the past 5 years (including South Orkney, South Georgia, Chagos, and Motu Motiro Hiva), and counting fisheries protected areas and other kinds of site protection, claimed that 'Rates of progress suggest that it may well be possible to attain 10 per cent coverage by 2020'. Moreover, as Jones (2014, p. 8) notes, 'only the IUCN target specifically requires *no-take* MPAs…The other targets, including under the CBD, could be met by partially protected MPAs that include provisions for some extractive activities' (italics in original). And if areas are counted where temporary (e.g. seasonal) protection is afforded, and/or real time restrictions are imposed to protect spawning or juveniles, the global targets may well be feasible.

Nevertheless, there are drawbacks with targets. One drawback is that the targets may have little scientific justification—'ecologically irrelevant' (Metcalfe et al. 2013, p. 10)—because they have been virtually picked out of the air. One conservation scientist, who has complied IUCN guidelines, explained that the 10 % figure that was initially adopted by the CBD was formulated arbitrarily over dinner: a figure of 5 % was suggested but it was thought that 5 % would be more likely to be achieved if a 10 % target were set (pers. comm). On the call for a 20 % global target, Agardy et al. (2003, p. 361, 359, 360) stated that 'The 20 % figure has been elevated to a dogmatic standard for a minimum proportion of a type of ecosystem that must be delineated as no-take MPA in order for the MPA to be effective in protecting natural resources', but it was originally 'extrapolated from very specific localized studies of particular fisheries within particular habitats—not from representative community ecology data from a wide variety of habitat types…Yet the 20 % figure has been adopted as the mantra of some MPA advocates targeting a wide range of objectives under a diverse spectrum of ecological and social conditions…Does the ease of convenience in selecting a single rule-of-thumb figure for all situations run the risk of selecting meaningless threshold targets?'. Similarly, Planes (2011, p. 327) reported that 'In the late 1990s a magic number came out of some modelling studies suggesting that 30 % of the coastlines should be included in MPAs in order to ensure maintenance of the overall ecosystem functioning'. The US Coral Reef Task Force recommended that a minimum of 20 % of the southern Atlantic coast of the USA should be included in MRs on the basis that it would protect 20 % of the spawning biomass, a threshold below which stocks were likely to collapse (Bohnsack et al. 2000). Studies have claimed that setting aside 10–30 % of the sea as MRs in regions dominated by fishing impacts, can improve overall 'ocean health' by reducing the total cumulative impact on the ecosystem by 15–20 % (Halpern et al. 2010). But little scientific proof has been provided for these quantitative predictions. Moreover, 'overall ecosystem functioning', 'ocean health' and 'total cumulative impact' are highly contested concepts that are invariably left undefined. Wood et al. (2008, p. 349) claimed that targets are politically not ecologically driven: 'Targets have historically been justified in terms of political expediency rather than ecological knowledge' (see also De Santo 2013). Targets do not necessarily synchronise with areas requiring protection. As both De Santo (2013) and Metcalfe et al. (2013) noted, percentage area targets falsely imply that biodiversity is evenly distributed across the seas, and that the marine conservation needs of a country coincide with a particular percentage of protected sea area.

Another drawback of targets is that even where the targets do have scientific justification they may require enormous areas of MRs, which are politically untenable. Kaiser (2005) pointed out, for example, that a study had shown that even excluding fishers from 25 % of the North Sea would have only a negligible effect in protecting cod spawning stock biomass, while another study proposed that such highly mobile species required MRs spanning up to 65 % of particular seas.

Targets may also create the illusion that if they are met, marine species will be guaranteed protection (Metcalfe et al. 2013; Dulvy 2013) whereas the truth is that many other measures need to be in place, the most important of which is enforcement of MR regulations. Without enforcement, targets may be met by purely 'paper parks' (Dudley 2008). An excessive focus on MPAs might deflect policy makers from introducing other measures which are also essential to protect fish stocks (Agardy et al. 2003). Many authors argue that MPAs are necessary but not sufficient for marine conservation (Allison et al. 1998), and must be used alongside other management measures to prevent declines in, for example, particular fish species (Blyth-Skyrme et al. 2006; Pastoors et al. 2000; Little et al. 2010; Fraser et al. 2009).

Targets encourage countries to meet them by designating MRs in areas where there is little fisheries activity, and therefore limited fisheries or biodiversity benefit. Dulvy (2013) said that targets led to the designation of 'super-sized' MPAs which added huge areas of protection, but which may have low conservation value because they are located in sparsely populated areas where there is little fishing activity. Conversely, the need to meet targets may lead countries to hastily establish MRs in sub-optimal locations, with unanticipated socio-economic and ecological impacts (Hilborn 2011).

Underlying these drawbacks is a contrast between the time-frames adopted by NPs and SCs: whilst the former stress the urgent need to meet deadlines and to designate MRs on the currently best available knowledge, accepting that this may compromise on quality (Wood et al. 2008), the latter argue for the designation of MPAs gradually on a site-by-site basis (Bloomfield et al. 2012), acknowledging that such an approach will not meet international deadlines and may allow the continued access of potentially damaging activities to vulnerable habitats, but claiming that the MPAs will be more sustainable in the long term.

On the positive side of the argument, targets may be politically useful (Ray 2004). Wood et al. (2008) claimed that targets helped to mobilise support for MR policy initiatives, while De Santo (2013) and Metcalfe et al. (2013) argued that targets generated political will and put international peer pressure on countries to reach them.

5.4.2 Skewed Focus of Literature

The recommendation of Willis et al. (2003a) for more empirical research on MRs has been realized, in that the proportion of empirical studies has increased relative to theoretical and review/note literature (though Goni et al. (2011) imply that little has changed). However, most empirical literature has focused on effects of MRs on reef-type habitats in tropical and warm temperate ecosystems (see Fig. 5.7a) rather than on soft sediment habitats in cold temperate ecosystems, and on fish (see Fig. 5.8) rather than on non-target species and habitats (Edgar 2011). In fact, the literature is skewed in six respects: (1) towards warm waters; (2) towards hard near-shore habitats; (3) towards invertebrates in temperate waters; (4) towards fish rather than biodiversity; (5) towards success stories; (6) and towards MRs rather than MUMPAs.

On (1), skewing towards warm waters, our literature review found that 310 papers had studied the ecological effects of MRs during the period 1970–2010, but the majority of these studies (228) focused on measuring the effects of MRs on tropical coral and warm temperate rocky reef fish assemblages. Therefore, although most of these studies showed, for example, increases in fish biomass within the MRs, they left a question mark over whether such results are generalisable to cold temperate marine ecosystems like those found in the UK. There is considerable controversy in the wider scientific community over how exactly MRs can be beneficial for non-reef-based commercial fisheries characteristic of cold temperate ecosystems (Hilborn et al. 2004). As Rogers et al. (2013, p. 176) put it, 'MPA science has seriously lagged behind MPA advocacy; although the number of empirical studies has increased, there remains a big gap in MPA science of cold temperate ecosystems'. Polunin et al. (2009, p. 6) pointed out that 'Many species are highly migratory, yet little is known about effects of MPAs on mobile finfish such as cod that are major targets in North Sea fishery conservation' (see also Sweeting and Polunin 2005). Caveen et al. (2012) suggested that it may have been assumed that inferences from warm water MR studies could be applied to cooler temperate and cold temperate waters MRs. Alternatively, the prominence of tropical and warm temperate MRs in the empirical literature (see Fig. 5.6) may be because the effects of MRs in cold temperate ecosystems and polar ecosystems are more difficult to assess. Many cold temperate fish species demonstrate extensive seasonal movement (Willis et al. 2003b); gene flow and connectivity are likely to be higher with the extended larval duration observed at higher latitude; and many life history characteristics, such as growth rate, age at maturity, longevity and maximum body size, are correlated with latitude (Blanck and Lamouroux 2007; Hutchings and Griffiths 2010; Sumpton and Jackson 2010). Another problem is the difficulty of making direct observation in cold temperate waters due to poor visibility (Polunin et al. 2009). Also, the fact that soft bottom communities are often found in deeper water or intertidal areas makes it more difficult to study them. Fish not associated with structure tend to move over wider areas, meaning soft sediment habitats need surveys over larger spatial scales (Rotherham et al. 2007). It is easier to show an effect of protection on a relatively sedentary invertebrate species associated with reefs than on more mobile fish species where a more intense (and costly) sampling effort will be needed to overcome high spatial and temporal variability in the fish assemblage (Rotherham et al. 2007). One reason why large seasonal closures (Dinmore et al. 2003) and partially protected areas (Frank et al. 2000; Murawski et al. 2000; Sweeting et al. 2009) are more common than MRs in cold temperate ecosystems is because of this mobility of exploited species (Shipp 2003). Some authors argue that MRs have been more often established in the tropics because it is relatively simple and cheap to manage an MR there rather than to enforce complicated restrictions on gear or impose elaborate effort and catch controls, as traditionally happens in many high latitude countries (Sale 2002; Shipp 2003).

Another important difference is the comparative speed of recovery of fish within temperate and tropical reserves. Whilst some authors have suggested that MRs always have rapid ecological effects (Halpern and Warner 2002; Halpern 2003), others are more wary, suggesting that responses to protection in temperate waters (e.g. Australian MRs) are often slow, complex and species specific (Barrett et al. 2007). Indeed, a recent meta-analysis of the seven Mediterranean and Lusitanian MRs stressed that management should adopt an extended timeframe (>30 years) to evaluate their fisheries' effects (Vandeperre et al. 2011). Caveen et al. (2012, p. 202) point out the chicken and egg problem, that until MRs are established in cold waters, their effectiveness cannot be tested: 'the available data supporting establishment of cold temperate MRs are weak, yet scientists can scarcely improve that knowledge until MRs are established in these areas'. On the other hand, Stewart et al. (2009, p. 244) claimed that 'there are adequate numbers of temperate marine reserves, with study results published in peer-reviewed journals, to justify a detailed, synthetic examination of their effects'; Claudet et al. (2011, p. 43) reported that 'temperate marine reserves…performed as well as or better than marine reserves in the tropics'; and Laffoley (2012) asserted that 'people say it is all about coral reefs and it won't work in temperate ecosystems waters. In actual fact the benefits have been shown to be greater in temperate waters, that's the science of MRs analysis'.

On (2), skewing towards hard near-shore habitats, although Lester et al. (2009) suggested that ecological effects of MRs were similar in tropical and temperate regions, their meta-analysis examined few highly mobile or migratory species, and the vast majority of the reserves they covered were protecting near-shore rocky or coral reef habitats. This illustrates the fact that scientific evidence supporting temperate MRs is strongest for hard complex near-shore habitats, and that 'data from soft sediment systems at temperate and tropical latitudes is severely lacking in MR science' (Caveen et al. 2012, p. 201). This is an important gap in knowledge, given that the majority of continental shelf seabed is soft sediment; for example, soft sediment covers 90 % of the shelf in the Antarctic, about 95 % of the GBRMPA, and virtually all of the proposed English North Sea Marine Conservation Zones network. A number of differences between soft sediment and reef-based systems may influence MR effects. Fish species associated with reefs are generally more site-attached (Barrett 1995; Zeller 1997; Tolimieri et al. 2009), as are individuals within species that range over both soft and hard bottoms (Attwood and Bennett 1994; Willis et al.

2001), and they are therefore likely to experience greater protection than those in sedimentary systems. This site attachment is a function of multiple behaviours, including territoriality (Barrett 1995), aggregation around structure (Grossman et al. 1997; Franks 2000), and predator avoidance. These differences may be the reason why less research has been conducted on soft-sediment systems—because of preconceptions that such habitats will not retain biomass, or because studies that have been conducted have not yielded statistically significant differences between protected and unprotected areas, and have therefore not been published (Edgar 2011). Of the few studies of MR effects on soft habitats in cold temperate waters that have taken place, some of them have given ambiguous results. For example, the recovery of benthic fauna was hailed as a successful result of the closure of the Lyme Bay reefs to mobile fishing gear, but Rodwell et al. (2013, pp. 256–257) pointed out that 'Of the 16 indicator species surveyed, only three showed clear recovery trends, and these were difficult to link unequivocally to the closure'. Advocacy for the use of MRs in soft sediment areas by many environmental organisations and some marine scientists thus appears to have limited empirical justification. That said, without the establishment of MRs in such habitats, the potential effectiveness of spatial protection measures in soft sediment systems cannot be measured[2]. It could be argued that the lack of MRs established over soft sediment bottoms in cold temperate ecosystems is the main reason why most studies on reserve effects are limited to fauna associated with reefs (Lester et al. 2009).

On (3), skewing towards invertebrates in temperate waters, whilst most studies of tropical or warm temperate MRs have concentrated on fish assemblages, most studies of cold temperate MRs have concentrated mainly on invertebrates (Caveen et al. 2012). On (4), skewing towards fish rather than biodiversity, Angulo-Valdes and Hatcher (2010, p. 637) reported that although more than half of the published studies on MRs were focused on fishery issues, reflecting the fact that the primary role of MRs across the world is to benefit fisheries, 'true fishery benefits from MPAs have very rarely been unequivocally demonstrated' (see also Stelzenmuller and Pinnegar 2011).

On (5), skewing towards success stories, Kaiser (2005, p. 1194) claimed that the scientific evidence used to justify MRs 'has been drawn from those studies that demonstrate a positive outcome of MPA implementation'. Kaiser criticised Halpern (2003)'s meta-analysis of MRs for concentrating on successful MPAs, and Halpern (2003, p. S118) himself admitted that 'Reserve success stories end up serving as the primary evidence for these assumptions [positive fisheries effects of MRs]… even though many examples exist where reserves did not provide the necessary functions'.

On (6), skewing toward MRs rather than MUMPAs, Rassweiller et al. (2012, pp. 11884, 11886) stated that studies of the ability of protected areas to increase fishery returns 'have focused on no-take MPAs and give us little indication of how

[2] In 2004, a UK government report entitled *Net Benefits* (2004) made the recommendation that to resolve this issue the fishing industry should engage with the conservation sector to do some large-scale no-take trials to see what the benefits were, but these trials were never undertaken.

a more flexible form of spatial management would impact profitability' (see also Jones 2007). On the other hand, Wenzel and Brock (2013, p. 1) claim that the scientific justification for MRs is slim because there are so few of them: 'The global rarity of no-take MPAs means that there is not always a great deal of scientific information available for different ecosystems and habitats where benefits to an adjacent fishery have been well documented'. In the USA, for instance, only 3 % of the total MPA area managed by NOAA is MR.

5.4.3 Mixed Evidence

The third theme is that the evidence in the literature for beneficial effects of MRs for fisheries is mixed. On the one hand, confident statements are often made that MRs enhance fishery stocks. For example, the European Marine Board declared that 'The fisheries management benefits of these ecological reserves are…increasingly recognised and include protecting critical feeding, nursery and spawning grounds which in turn help to build and maintain fish populations resulting in improved fishing yields' (Olsen et al. 2013, p. 5). There are many theoretical or modelling studies that purport to demonstrate the fisheries benefits of MRs. For instance, Pitchford et al. (2007, p. 286) reported that 'Compared with harvest control rules based on uncertain estimates of stock size, our simulations indicate that MPAs can substantially reduce the risk of fisheries collapse for only a very small cost to total yield'. Several meta-studies have also claimed that MRs have been successful in enhancing fish stocks. For example, Edgar et al. (2014, p. 216), in a study of 87 MPAs, found that 'Effective MPAs…had twice as many large (>250 mm total length) fish species per transect, five times more large fish biomass, and fourteen times more shark biomass than fished areas'. Similarly, Halpern (2003)'s review of 112 assessments of 80 MRs found that all biological values were much greater inside the reserves than in reference sites: population densities were 91 % greater; biomass was 192 % greater; and size and diversity were 20–30 % greater (see also Côté et al.'s (2001) meta-analysis of published data on 19 MRs).

On the other hand, Beare et al. (2013, p. 50) claimed that 'Empirical studies evaluating the performance of MPAs…are limited', and some of the empirical studies that do exist produced uneven results. Jones (2006, p. 146) noted that 'NTZs may actually undermine fisheries conservation objectives as a result, for example, of increases in fish stock predators and of lower growth rates due to overcrowding'. Sweeting and Polunin (2005, p. 2) stated that 'Evidence for benefits to temperate finfish inside MPAs is inconsistent'. O'Sullivan and Emmerson (2011, p. 115) reported variable findings in Europe's oldest MR—Lough Hyne—designated in 1981: 'Marine reserve designation has led to an increase in predatory crabs and *M. glacialis* [starfish], a subsequent decrease in primary consumers, especially the sea urchin, and an increase in macroalgal cover which is indicative of a trophic cascade. The study shows that establishing a Marine Reserve does not guarantee that conservation benefits will be distributed equally…there may be unforeseen consequences of MR designation'. Although Lester and Halpern (2008, p. 53) reported that their synthesis study of 20 papers published during 1977–2007 on 21 MRs in 11 countries found that 'no-take areas had higher biomass, density, species richness and individual organism size on average relative to partially protected areas', they admitted that 'This difference was statistically significant only for density', and that there was 'considerable variability in the documented effects of no-take versus partial protection, likely resulting from various factors that could not be accounted for in our analyses due to lack of information'. Moreover, Graham et al. (2011, pp. 112–113) pointed out that although 'Meta-analyses found that density and biomass of reef fishes were higher in NTAs compared to fished areas…particularly species targeted by fishing…Such differences may often be the result of…the "file drawer syndrome", whereby null results are rarely published and are therefore unavailable to meta-analyses…[Moreover] such studies…rarely account for the possibility that the NTAs were initially sited in areas of high fish abundance and biomass'.

There are in fact 11 sources of controversy in this third theme of mixed evidence: the spillover effect; larval export; the displacement effect; habitat effects; trade-offs between biodiversity and fisheries; large-scale MRs; MR networks; the role of science; timeframes; socio-economic effects; and political controversy. Data from studies of the spillover effect are particularly problematic. Many writers confidently affirm that evidence of spillover is robust. For example, Metcalfe (2013, p. 8) wrote that 'Evidence supporting the contribution of spillover to exploited areas has been demonstrated by several studies that have indicated…increased catches per unit effort and increased population sizes in adjacent areas…and harvests of larger and more highly valued species'. Russ et al. (2003, p. 18, 19) stated that their study of a marine reserve at Apo Island in the Philippines 'presents some of the most convincing evidence for spillover to date' (see also Alcala and Russ 2006), and that 'Several other studies have produced convincing evidence in support of spillover from coral reef and temperate marine reserves'. Christie et al. (2010, p. 1) affirmed that 'there is mounting evidence for localised spillover'.

However, Agardy et al. (2011, p. 229) stated that 'Spillover has not been demonstrated to the same degree as increased production inside reserves'; Stewart et al. (2008, p. 23) reported that 'the available evidence is insufficient to evaluate the effectiveness of temperate no-take zones for maintaining sustainable fisheries through the provision of overspill effects'; Gaines et al. (2010a, p. 18251) claimed that 'The effects of marine reserves on adjacent fisheries are far less clear than the potential conservation benefits'; Russ et al. (2003, p. 15) affirmed that 'No study has unequivocally demonstrated spillover…from a marine reserve' (see also Russ and Acala 1996); Sale et al. (2005, p. 77) asserted that 'we have remarkably few well designed studies of no-take reserves that can rigorously demonstrate that they have sustained or enhanced fishery yields in the surrounding region. Solid evidence of recruitment subsidy does not yet exist, and much of the evidence of spillover is equivocal'; Rakitin and Kramer (1996, p. 111) reported in their study of the Barbados Marine Reserve (BMR) that 'The evidence we found for emigration of fish from the BMR was limited and inconsistent'; Rogers et al. (2013, p. 176) pointed out that 'The evidence base for beneficial fishery impacts for MPAs in other ecosystems where species are mobile over considerable distances, such as temperate and tropical continental shelves, is very weak'; and Kaiser (2009) claimed that 'there is no evidence of spillover from studies done to date, and furthermore most studies have been so badly designed that it is almost impossible to disentangle the effect of the NTZ from a confounding habitat effect'.

There is some evidence for the net movement of fish from inside cold temperate MRs to fished areas outside (Cole et al. 2000; Fisher and Frank 2002), and a few studies suggest that MRs have the potential to increase CPUE (Guidetti et al. 2010; Murawski et al. 2005), but no cold temperate studies have shown MRs to increase fisheries yields through spillover. The only evidence for increased yields from spillover comes from a handful of studies of tropical and warm temperate MRs (Rakitin and Kramer 1996; Russ et al. 2004; Ashworth and Ormond 2005). Some argue that the 'fishing-the-line' phenomenon is an indication that catches are greater next to MR boundaries (Dan Laffoley, pers. comm.), and there is some empirical evidence to suggest that this may be true in certain cases (Murawski et al. 2005). Detecting spillover effects requires methodologies that are expensive and not straightforward to implement (Sale 2002). There are very few spillover studies that have long-term monitoring data, which is essential to inform strategies for sustainable development (Dan Laffoley, pers. comm.; Agardy 2010). Evidence of the spillover of adult fish across MR boundaries that compensates the catches of local fishermen is generally limited to small-scale MRs (Vandeperre et al. 2011).

Providing definitive evidence for larval export from MRs is even more difficult (Russ 2002; Graham et al. 2011; Gell and Roberts 2003). Sweeting and Polunin (2005, p. 2, 50) stated that 'data on it are rare, even for the well-studied tropical reef MPAs. The magnitude of the larval export role of MPAs can therefore scarcely be predicted…empirical evidence of larval export is suggestive or circumstantial rather than definitive. Evidence for finfish larval export is virtually absent'. Gaines et al. (2010a, p. 18253) stated that 'we know very little about marine reserves as sources and sinks of larvae, even after over a decade of attempts to quantify the origins of larval supply' (see also Hamilton et al. 2010). There is some circumstantial evidence to suggest that larval export does occur (Beukers-Stewart et al. 2005; Roberts 2003), including direct measurement of enhanced larval recruitment downstream of a MR (Cudney-Bueno et al. 2009), but increased larval production due to an increase in spawning stock biomass within the reserve is often simply inferred (Beukers-Stewart et al. 2005) since to measure such an effect directly is very difficult[3] (Hedgecock et al. 2007). As Gaines et al. (2010b, p. 18291) pointed out, 'the results are species-specific and difficult to quantify accurately'. Pelc et al. (2010, p. 18266) explained that 'Inherent difficulties in collecting and interpreting the data needed to document larval export may explain the relative scarcity of empirical evidence. Benefits from larval export may be widespread but very diffuse, and the signal of export may be too weak to detect relative to the high spatial and temporal

[3] Though there are some studies that have tried to do this (Cudney-Bueno et al. 2009; Christie et al. 2010).

variability of recruitment'. Christie et al. (2010) claimed to have overcome these difficulties by providing genetic evidence of larval export within an MPA network in Hawaii. Perhaps the more important question is not whether larval export occurs, but 'whether it provides a benefit to fished areas sufficient to outweigh the increased mortality that may result from displaced effort when closing an area to fishing' (Pelc et al. 2010, p. 18266).

Displacement of fishing effort from an MR to another area may (or may not) undermine the fisheries benefit of the MR, as Jennings (2009) noted. Data on displacement effects of MRs are very sparse: Hilborn et al. (2004, p. 201, 202) stated that displacement 'may have a number of undesirable consequences that in most cases remain un-analysed'. Likewise, Agardy et al. (2011, p. 229) asserted that 'Displacement…is poorly defined and few empirical studies have quantified the impacts of fisheries closures that cause displacement in fisheries'. Suuronen et al. (2010, p. 237, 238) reported evidence of negative MPA displacement effects in the Baltic cod fishery: 'The enlargement of Bornholm MPA in 2006 caused substantial effort displacement towards areas dominated by smaller sized fish. This contributed to the increased discarding of juvenile cod…These findings on MPAs poor or negligible protection potential are well in line with other studies on the effects of MPAs on mobile fish species'. Likewise, Sweeting and Polunin (2005, p. 54) noted that 'Closed areas for the protection of cod in the North Sea altered effort distribution such that the impact on long lived benthic species vulnerable to fishing was greater'.

Displacement was the single most important issue for the anti-MR MPA Coalition (MPAC) in the UK (as we shall see in the next chapter). The MPAC's chairman, Lockwood (2013a), stated that 'throughout the whole process of establishing a network of MPAs our primary concern has been the potential for displacement of fishing activity from its customary areas, with adverse socio-economic and ecological consequences'. He claimed that 'NE say displacement is not their concern because it's a socio-economic issue, but we claim it's an ecosystem matter as well—if boats must move off a designated site they will have to go elsewhere, resulting in increased environmental pressure at the new site' (*Fishing News* 25.11.11: 8). On the other hand, Ballantine (2014, p. 6) argued forcefully from the NP perspective that the displacement problem was a purely social construction: 'This is a classic example of an 'invented problem'…fisheries…'displace' all the time for a variety of reasons (e.g. market forces and fashions, fuel costs, tax and subsidy arrangements)…marine reserves will cause the 'displacement' of many human activities, including coastal development, tourist destinations, outdoor education and many forms of recreation. No useful…allowance can be made for them. The sensible reaction of the authorities to the idea of displaced fishing is simply to ignore it'.

A serious problem for assessments of the benefits of MRs is that reserve effects may be confused with habitat effects: 'Habitat differences between NTMR and fished sites may confound analyses of NTMR effects, even when attempts were made to ensure sites were as similar as possible. Many NTMR studies have not accounted for potential confounding effects of habitats on marine assemblages, resulting in ambiguous conclusions about NTMR effectiveness and utility. Proponents of NTMRs may unintentionally overstate reserve benefits, and critics may claim that benefits are uncertain due to poor study design' (Miller and Russ 2014, p. 52). In their review of 164 MR studies, Miller and Russ (2014, p. 51) found that 'Over half of the studies (54.3 %) made no statistical attempt to account for habitat effects'. Moreover Miller and Russ (2014, p. 56) reported that when some studies statistically corrected their findings for habitat effect, they found that 'fewer species exhibited significantly higher densities inside reserves than outside'. A solution to this problem would be to conduct BACI analyses (or, preferably, before-during-after-control-impact (BDACI) analyses (Miller and Russ 2014)), though this may not always be feasible, practically or financially.

The issue of trade-offs between biodiversity and fisheries benefits of MRs aroused controversy in the literature. For some writers, there is no trade-off, since MRs benefit both biodiversity and fish stocks—that is, a win-win situation. For example, Rife et al. (2013, p. 200, 201) baldly stated that 'Although primarily used to protect marine habitats, vulnerable species, and ecological processes from destructive human activities, marine protected areas (MPAs) can simultaneously enhance fisheries stocks via larval and adult spillover', yet they admitted that 'clear examples of success…are rare. Research findings instead show that MPAs have generally been unsuccessful in meeting their goals; failures far outnumber the successes'. They attributed the failures to factors such as inadequate funding, poor enforcement, too few NTZs, and lack of community support, but if there are so many failures, the question arises whether the assumption that MRs can protect both biodiversity and commercial fisheries is sound. For other writers, there is always a trade-off between these two objectives. Jones (2006, p. 146) stated that 'it is debatable whether NTZs can be pursued as a single tool to achieve both aims. NTZs aimed primarily at achieving marine biodiversity objectives will have different design and implementation criteria to those aimed primarily at achieving fisheries management objectives' (see also Brander 2010). The claims of win–win or double benefit have come from scientific studies of mainly small-scale MRs (Stelzenmuller et al. 2009), and may not be applicable at a larger scale (Greenstreet et al. 2009).

Large-scale (LS) MRs also provoked evidential controversy. De Santo (2013) pointed out that during 2000–2010 there was a rapid expansion in the designations of LSMRs (see also Jones 2011). This was partly because the threat of climate change led to calls to protect as much of the seas as possible to safeguard marine features such as coral reefs which were highly vulnerable to global warming; partly because the international community was pressing for protected area targets to be met on time; partly because of the need to protect highly mobile fish species (Hilborn et al. (2004, p. 202) pointed out that 'Many of the species caught in industrialized and some artisanal fisheries are so mobile that marine reserves would have to be very large to effectively protect breeding stock' (see also Koldewey et al. 2010; Sweeting and Polunin 2005); and partly because LSMRs were more resilient than small MRs. LSMRs would seem to yield greater benefits than small-scale MRs because, as De Santo (2013, p. 143) explained, bigger areas provide more space for spillover and larval dispersal to take place, and 'The larger the area, the smaller its border-to-area ratio, reducing the amount of 'edge' habitat (i.e. habitat close to the edge of a protected area is more exposed to outside pressures/threats than habitat located closer to the centre)' (see also Jones 2011). Lubchenco et al. (2007, p. 14) claimed that 'a bigger marine reserve can protect more habitat types, more habitat area, bigger populations of animals, and a larger fraction of the total number of species in an ecosystem. Bigger populations in areas with more species are especially important as insurance against catastrophes, such as hurricanes or oil spills'. However, the empirical evidence for such benefits was slim. For one thing, as Sale et al. (2005, p. 75, 77) pointed out, 'reserves cannot be simultaneously of optimum size for all contained species, and widely ranging rare species might never be adequately conserved using reserves…some species might be too mobile for management using reserves to be practical. It might not be politically possible to implement reserves of sufficient size to provide them with the level of protection required'. For another thing, 'there is no…evidence showing that extremely enormous MPAs on the scale of hundreds of thousands of kilometres are more likely to achieve conservation objectives than effectively managed networks of MPAs. Indeed, extremely enormous MPAs run the risk of being no more than 'paper parks' due to the difficulty of monitoring and enforcing them' (De Santo 2013, p. 138; see also Murawski et al. 2005; Olsen et al. 2013). Also, as time goes on, it will become increasingly difficult to find sufficiently large areas for LSMRs (Jones 2011).

MR networks raise several data issues. Grorud-Colvert et al. (2011, p. 295) drew attention to the lack of empirical data on the effect of MR networks, and the lack of an 'established framework for assessing whether ecological effects across the network as a whole are greater than the sum of the effects within the individual reserves in the network' (see also Gaines et al. 2010a). Jones (2014, p. 28) rehearsed the enormous difficulty of providing data to justify the establishment of 'ecologically coherent' MPA networks—i.e. 'a detailed understanding of the dynamic patterns of currents, larvae dispersal routes, fish migration routes, genetic population structure, etc, in a given region'. This difficulty has led some proponents of MPA networks to adopt a more practical approach, by using criteria such as representativity and replication criteria to achieve coherence (Jones 2014). Another practical approach is to select surrogates for ecosystem health such as substrates (Banks and Skilleter 2010). But critics regard representativity, replication, and surrogacy as highly speculative substitutes for hard evidence. On representativity, for example, Rice and Houston (2011, p. 650) referred to 'irreconcilable debates between resource users who consider a region "represented" once a few tens of km^2 have been protected, and conservation advocates who argue nothing less than protection of a large, intact exemplar of a given biogeographic type is adequate to "represent" the type'. Connectivity was another problem faced by MPA networks, because MPAs need to be close enough to benefit from mutual colonisation, but not so close that they could succumb to the same catastrophic event (Wagner et al. 2007). In any case, as Carson and Hentschel (2006, p. 111) pointed out, 'conservationists must recognise that a regional network of MPAs will never protect all species'.

The role of science in the MR literature was another controversial issue in the literature, in that many advocates insist that the decision about MRs should be purely scientific, but there are significant limits to the scientific knowledge available.

There is a growing literature documenting the use of sophisticated scientific models in the planning of MR networks (see Fig. 5.11). Decision support tools such as MARXAN (Ball and Possingham 2000) have been developed to inform policy makers where the optimal placement of MRs should be to meet defined fisheries and conservation objectives and minimise costs to stakeholders (Klein et al. 2008a, b). These tools are being used extensively to inform the design of MR networks in the USA (Klein et al. 2008a, b; Ban 2009) and Australia (Game et al. 2008). However, as noted above, the data needed for such a process are vast[4], requiring spatial information on habitat, species distributions, larval, juvenile, and adult movements and source–sink dynamics of larval production and recruitment (Jones and Carpenter 2009), as well as spatially explicit socio-economic data (Bloomfield et al. 2012). Some scientists argue that the paucity of scientific data is not a sufficient reason for declining to designate MRs (Abdulla et al. 2009), and justify this argument by invoking the precautionary principle (PP). However, critics of MRs claim that invoking the PP is a tacit admission that the decisions are based not on science but on normative judgements, and that if MR decisions are normative, all stakeholders, not only scientists, should share in contributing to them. In other words, MR decision making is not just about the science (the linear model): it requires an inclusive stakeholder process that encourages people with different perspectives and interests to engage with one another to navigate trade-offs over the use of the marine environment (the deliberative model) (Salomon et al. 2011). Indeed, some would argue that protection of the marine environment is essentially a societal (i.e. normative), rather than a scientific (i.e. empirical) decision, in that establishing MRs to protect nature is a statement of what society judges to be the right thing to do. If the general public wants more sea to be protected as MRs, then who is to say that this is wrong? Worryingly, Connell (2013, p. 84) noted that 'Most MPAs…have been developed without any detailed assessment of the value of such areas to local communities' (see also Olsen et al. 2013; Charles and Wilson 2009).

Timeframes raise another evidential issue: how long do we have to wait before we can pronounce an MR successful? Babcock et al. (2010, p. 18256) noted in their decadal-scale study of MRs that 'Most target species showed initial direct effects, but their trajectories over time were highly variable. Many target species continued to increase, some levelled off, and others decreased'. One respondent in our questionnaire survey of scientists on bias said that MRs worked, but not instantaneously: '*MNRs obviously work…The problem is unwarranted claims about the speed at which they work—very rapid responses and things that MNRs trigger; trophic cascades and increased 'resilience', whatever that may mean…The appreciation of MNRs is reduced as people make unsubstantiated claims about their efficacy*'.

On socio-economic effects of MRs, many researchers drew attention to the lack of data. For example, Sweeting and Polunin (2005, p. 54) stated that 'MPAs have socio-economic consequences about which very little is known or acknowledged despite MPAs being assumed to create net social and economic benefits'. Rudd

[4] In data poor situations the use of MARXAN should be avoided as output will be meaningless (Jeff Ardron, pers. comm.).

et al. (2003, p. 80) claimed that 'the costs of marine reserves to fishers have been poorly quantified to date and there has been insufficient consideration of fisher behaviour'. Richardson et al. (2006b, p. 1192) pointed out that 'Socioeconomic issues...seldom receive consideration concomitantly with ecological issues'. Olsen et al. (2013, p. 51) asserted that 'Empirical assessment of socioeconomic effects of MPAs is sparse' (see also Edwards et al. 2009; Gaines et al. 2010b). Graham et al. (2011, p. 120) stated that 'Few studies...have examined the effects of NTAs on aspects of well-being such as community values, poverty, human health and empowerment'. Christie (2011, p. 177, 180) claimed that 'Advocating for global MPA networks commonly overlooks various important social considerations. How MPAs affect various and complex human communities and the level of acceptance of MPAs by a diverse society is little documented and examined...Only few and limited comparative studies of the human and management dimensions of MPAs... exist'. Sale et al. (2005, p. 77) argued that 'The socioeconomics of no-take reserve introduction are not yet well understood'.

Yet several writers claim that MRs cause serious short-term harm to fishers and their communities: fishers have to travel further to unfamiliar grounds which entails higher fuel costs; there is greater risk to the lives of crews on smaller vessels (Hannesson 1998); and the resulting higher prices for fish (Pitcher and Lam 2010) could put out of business fishers who were making marginal profits (NFFO 2009b). Graham (2011, p. 121) wrote that 'The establishment of NTAs can...marginalize local users by exacerbating conflicts over coastal resources, curbing ownership and use rights, and reducing participation in decision making'. Richardson et al. (2006b, p. 1201) claimed that 'In terms of present economic value, short-term fishery losses will generally outweigh future gains in most systems'. Sweeting and Polunin (2005, p. 2) stated that 'In no case examined has spillover compensated for loss of fishing area'. Moreover, as Charles and Wilson (2009) pointed out, there may be unfairness in the way the impacts of MRs fall on different groups of people. Richardson et al. (2006b, p. 1200) stated that 'we believe that detailed socioeconomic surveys are as important to the design of successful cost-effective marine-reserve networks as detailed biodiversity data', while Sale et al. (2005, p. 77) warned that 'Without attention to the underlying socioeconomic issues, science-based reserve development will be significantly constrained, and is unlikely to serve scientific or other needs effectively'.

On political controversy, the claim that MRs lead to less political conflict than does CFM, is dubious. Sweeting and Polunin (2005, p. 50, 2) reject the argument that whereas CFM is blighted by political resistance to scientific advice, MRs are not: 'It has been suggested that MPAs may reduce political interference in fishery management, because politicians may be less willing to reopen no-take MPAs than they are to overrule other fisheries management measures. However...Size, position, configuration and level of protection may all be politically modified...The argument that MPAs are more politically robust than other forms of fisheries management is rejected. The notion that MPAs reduce conflicts among users is valid in some cases but not others'. We will see in the next chapter how considerable political conflict was generated by proposed MPAs in the UK.

All 11 issues of mixed evidence illustrate two general problems with the literature on MRs—data deficiency; and species-specificity. On data deficiency, Sweeting and Polunin (2005, p. 43) claimed that 'Common to virtually all studies is a lack of rigorous science in the determination of the positive and negative effects of MPAs'—mainly because of lack of BACI analysis: 'Without BACI approaches, effects of MPAs can only be inferred rather than attributed'. Stewart et al. (2008, p. 11) noted in their meta-analysis of MPA studies 'the existence of large knowledge gaps in the evidence base. No studies were retrieved for any no-take areas in sand or mud habitats…[nor] concerning commercially important, highly migratory, northern hemisphere fish species, and the data regarding pelagic fish species are based on only two no-take areas'. Jones (2006) distinguished between data for fisheries benefits, which he claimed were sparse, and data for biodiversity benefits, which he claimed were plentiful. However, Peckett et al. (2014, p. 334, 338) stated that data deficiencies on biodiversity were vast: 'comprehensive biological data, such as presence/absence data for species or detailed habitat maps, are not available for the majority of the marine and coastal environment'. Stewart et al. (2009, p. 8) claimed that 'Our systematic review has revealed clear gaps in the evidence base regarding the effectiveness of temperate marine reserves for either biodiversity conservation or sustainable fisheries management'. On data deficiency of fisheries effects, Jones (2014) suggested that this deficiency is because of the paucity of MRs where research can be conducted (see also Mora 2011).

Part of this difficulty of assessment lies in distinguishing between the effects of MRs and wider environmental factors (Jones 2014). Stewart et al. (2008, p. 4, 5) pointed out that even if fish density is higher in MRs than in adjacent areas, 'it is not clear if these differences are due to marine reserve effects or other differences between the marine reserves and comparator, such as habitat variation' This difficulty is due to inadequate analysis of habitat effects on data: only 20 % of empirical studies have measured habitat and included this as a co-variable in their analysis (see Fig. 5.10b). Effect magnitude may be overestimated if an MR is located in more productive habitat (Hilborn 2002), or underestimated if illegal fishing still occurred in the MR (Guidetti et al. 2008). Few studies have tried to quantify fishing effort occurring inside (i.e. through illegal fishing) and outside MRs (Bloomfield et al. 2012), yet the magnitude of fishing effort outside the MR and inside before designation will play a key role in determining the direction and magnitude of the reserve response (Lester et al. 2009). Recent evidence suggests that newly established MRs are often located in resource-poor areas due to socio-political factors, and when surveyed, they have significantly fewer fish than nearby control locations (Edgar et al. 2009; see Table 5.2). Another part of the difficulty of assessment is that, despite calls for more rigour in experimental design (e.g. spatial and temporal replication) when empirically assessing MR effects (Guidetti 2002; Willis et al. 2003a), a high proportion of empirical studies are snap-shot (see Fig. 5.10a) with 25 % of these studies being spatially confounded by using only one fished control area. Very few studies have implemented a fully-replicated BACI design (Underwood 1993), though there are some moves in this direction, for example, through the application of fully-replicated asymmetric monitoring (Hoskin et al. 2010). The result is that, as Hamilton et al. (2010, p. 18275) stated, 'Networks of relatively large MPAs are being implemented in many locations across the globe, but few networks have been evaluated'. This may be the main reason why Dulvy (2013, p. 359) claimed that 'there are surprisingly few clear examples of MPA success'. Sweeting and Polunin (2005, p. 56) observed that 'Appeals for measures such as MPAs to address the shortcomings of fishery science have to face up to the greater inadequacy of MPA science'.

On species-specificity, one respondent in our scientists' questionnaire said that MRs worked for some species but not for others: '*Some of the papers I have submitted have shown positive and others negative and others no effects at all of MNRs on different species. The direction of the effect, or whether or not any effect is actually detected at all, depends on the position of the species within the complex web of interactions making up the marine ecosystem of interest, and how the species being analysed either clearly interacts (or not) with species that are targeted by fishers.*' Stewart et al. (2008, p. 18, 19, 20) reported on their meta-analysis of MPAs that 'No-take zones did not have significantly increased densities of pelagic fish species…no-take areas did not show significantly increased densities of highly commercial species…No-take areas did not significantly increase the density of species that inhabit artificial reefs or breakwaters'. Halpern and Warner (2002, p. 365) noted that MRs affect different species in different ways: 'results for a particular species will certainly depend on their life-histories…For example, after massive reserve closures in the Georges Bank area…cod stocks have been slower to respond to protection, whereas scallop populations quickly grew to enormous size'. Even NPs such as Lubchenco et al. (2007, p. 5) admitted that a high proportion of fish species do not thrive in MRs: 'A worldwide analysis found that 61 % of fish species were more abundant inside reserves than outside, whilst 39 % of species were more abundant outside reserves than inside…Some fish and invertebrate species become less abundant in an area after it is designated as a marine reserve. Such declines generally reflect interactions among species, such as larger numbers of a predator eating more of its prey'. In other words, some species within a MR might not increase because of trophic cascades: 'The lack of fishing inside the reserve might lead, through processes such as trophic cascades, to changes in community structure that cannot currently be predicted explicitly. The rule that populations of fishery species will be more abundant, larger, older and therefore more fecund inside a reserve might not hold if such shifts in community structure occur' (Sale et al. 2005, p. 77; see also Sweeting and Polunin 2005). Stokesbury et al. (2007) reported a mass mortality of sea scallops during the mid-2000s within the Nantucket Lightship Closed Area, an MPA in the Great South Channel, USA, because of senescence and parasitism. Exploited species respond more quickly, extensively, and persistently to restrictions than do non-exploited species: Hart (2006, p. 1449) stated that 'long-term closures are unlikely to increase yield when open area fishing mortality is at or less than that which produces MSY…On the other hand, closure can increase yield at high fishing mortalities, prevent stock collapse, and have other conservation benefits (e.g., for biodiversity)'.

Moreover, there seems to be a systemic difference between the rates of MR success in coral reef systems where fish are sedentary, and in temperate systems where fish are more mobile (Côté et al. 2001). Pastoors et al. (2000, p. 1014) stated that 'Empirical evidence for positive effects of MPAs is…largely limited to tropical reef systems…Evidence for the applicability of MPAs in temperate systems is scarce'. Jones (2014, p. 27), however, has disputed this finding: 'these studies include both tropical and temperate fisheries, confounding claims that such 'benefits beyond boundaries' are confined to…tropical coral reefs'. Moreover, there is substantial evidence of positive effects of cold temperature MRs on invertebrates, particularly shellfish and lobsters (Rogers-Bennett and Pearse 2001; Rowe 2002; Jamieson 2000; Hoskin et al. 2010). For example, no-trawl areas have been used to protect scallop grounds on Georges Bank (Murawski et al. 2000) and around the Isle of Man, UK (Beukers-Stewart et al. 2005) with considerable success, and scallop fishermen are now experiencing at first-hand the benefits that closed areas can make to their catches (Beukers-Stewart, pers comm.). After closures on the Georges Bank, the total scallop biomass increased by a factor of 14, and harvestable biomass increased by a factor of 15 over a 4-year period (Murawski et al. 2000). In the Isle of Man, the exploitable biomass of scallops increased by a factor of 11 over approximately a 14-year period, with circumstantial evidence for larval spillover from the closed area (with scallop densities increasing in adjacent fishing grounds), and anecdotal evidence for the spill-over of adults over the closed area boundaries (Beukers-Stewart et al. 2005).

So it seems that the fisheries effects of MPAs are highly dependent on scale and local ecological conditions, making it very difficult to generalise from one case-study to another (Holland 2000; Bloomfield et al. 2012). Graham et al. (2011, p. 117) stated that 'Catches or CPUE are generally higher in boundary areas compared with more distant areas, but results vary by species and habitat'. Similarly, Gaines et al. (2010a, p. 18253) pointed out that it depends on the species: 'whereas relatively sedentary species repeatedly show positive responses to reserve protection, the utility of marine reserves for highly mobile species, which may spend the majority of their time outside reserves, is often deemed negligible, although notable exceptions exist' (see also Edwards et al. 2009).

A questionnaire respondent said that MRs worked for some species but not for others:

> *Some of the papers I have submitted have shown positive and others negative and others no effects at all of NMRs on different species. The direction of the effect, or whether or not any effect is actually detected at all, depends on the position of the species within the complex web of interactions making up the marine ecosystem of interest, and how the species being analysed either clearly interacts (or not) with species that are targeted by fishers.*

Species-specificity is matched by habitat-specificity, and both imply site-specificity—that is, that benefits of MRs very much depend on where they are located. There are two contrasting theories of site specificity: (1) that MRs are not site specific, but everywhere generate benefits (e.g. Halpern 2003); versus (2) that any benefits generated by MRs are site specific (Osenberg et al. 2011). The first view implies that MRs necessarily have more benefits than less-restrictive MUMPAs. But whilst this may be true for heavily exploited fish species, for other aspects of biodiversity, MRs may have negative impacts through reverse trophic cascades as we have seen (Hoskin et al. 2010; O'Sullivan and Emmerson 2011). Ecological evidence about MRs that the first view relies on has come from very specific localities, and their benefits to fisheries are still very much debated and highly dependent on their contexts (Bene and Tewfik 2003; Fanshawe et al. 2003). Whether an MR is the right tool to meet its objectives, therefore, depends on circumstances—which is what the second view claims. Adherents of the second view argue that generalisations about the effects of MRs are unhelpful for policy as one cannot use such information to predict the time-scales and extent of recovery at the species/habitat level (Edgar and Stuart-Smith 2009).

The prevailing consensus in the literature review seems to be that MRs can improve yields in fisheries that have been recruitment-overfished and sometimes growth-overfished (Murawski et al. 2000), though it has been argued that such an objective could also be met by a reduction in fishing effort (Hilborn et al. 2004). Holland and Stokes (2006, p. 1183) challenged the findings of Rodwell and Roberts (2004) that MRs necessarily achieved higher rates of biomass than open fishing areas by pointing out that 'Although there is probably consensus that MRs may help conserve stocks and possibly increase yields in overexploited fisheries, neither empirical studies nor models have shown that they will benefit well-managed fisheries'.

5.4.4 Difficulties of Enforcement

The fourth issue that emerged from the literature is the difficulty of enforcing MR regulations. Many writers have pointed out that conventional fisheries management (CFM) often fails because of lack of implementation, but MR regulations are subject to the same hazard. A population model used by Le Quesne (2009, p. 132) showed that 'biomass and yield are reduced, and can collapse under 'bad' MPA management'. Similarly, Jennings (2009, p. 19) noted that 'In many MPAs, unauthorised fishing has compromised the achievement of objectives, and enforcement and compliance have been inadequate or unsupported'. It is claimed that one of the advantages of MRs over CFM measures is that they are easier to enforce (McClanahan and Mangi 2000). For example, it is claimed that detecting whether a vessel is illegally fishing in a protected area is simpler than detecting whether a vessel has a valid licence, is using legal gear, is within quota, and is adhering to minimum landing size (MLS) (Jones 2006). But where MRs are vast in size and/or located in remote areas or on the high seas, detection of illegal, unreported, and unregulated fishing (IUU) fishing may be almost impossible. It is telling that Lubchenco, Head of NOAA, confessed that 'We don't have the resources that we need to actually monitor, enforce and understand these areas' (quoted in Cressey 2011a). Reports of widespread non-compliance with MR regulations are common. In a review of coral reef MRs across the world, Mora et al. (2006) showed that medium to high levels of poaching occurred in 65% of them. In such circumstances, MRs may be little more than 'paper parks' (Monbiot 2012). Dunne et al. (2014) gave the example of Chagos, which they claimed is worse off with its current MR than with its previous CFM, because it faces exactly the same problems of IUU fishing, but has lost its fishers' 'eyes and ears' to report infringements. Rudd et al. (2003, p. 65, 77) asserted that a precondition of a successful MR is a high level of social and political capital: 'Reserves may be the most efficient policy option when both community and state capacity is high, but may not be when one and/or the other is weak'. But where there is high social capital, CFM could be equally effective in achieving fisheries management goals.

5.5 Conclusion

This chapter has examined the evidence in the peer-reviewed literature from 1970–2010 for the benefits of MRs. The evidence that MRs locally increase abundances of some susceptible species, and in that sense enhance biodiversity, is strong, but the evidence that MRs benefit fisheries is weak, while there has been very little research into the socio-economic benefits of MRs. We found that 87% of the empirical studies focused on MRs rather than MUMPAs; that 84% of studies concentrated on tropical/warm coral reef systems; and that most studies were snapshots rather than longitudinal studies, with few control sites or before-and-after analyses. Four major themes emerged from the results of this examination of MR literature. First, the drawbacks of targets like IUCN's aim for 20% of global seas to be designated as MRs by 2020: such targets had little scientific justification; encouraged over-hasty designations that were in the wrong places; unnecessarily threatened fishers' livelihoods; took resources away from more urgent causes; and created the false impression of problem solved. Second, the literature was skewed in favour of tropical/warm water and near shore locations; fisheries not biodiversity effects of MRs; success stories; and MRs rather than MUMPAs. Third, there was mixed evidence for the claims that MRs enhanced fish stocks by spillover or larval export; and that large-scale MRs and MR networks delivered significant fisheries benefits. There was a widespread recognition of the inadequacy of the scientific data available, and a growing realisation that MRs were species-specific, habitat-specific and therefore site-specific. Fourth, the problems of enforcement of regulations were no less real in MRs than in conventional fisheries management (CFM) regimes.

The next chapter is a case study which tests whether the NP discourse or the SC discourse was the main driver during the passage of the UK's Marine and Coastal Access Act (MCAA) in 2009, and used in the subsequent planning of a network of Marine Conservation Zones (MCZs). Many of the themes discussed in Chap. 5 will surface in Chap. 6.

Chapter 6
Case Study of the 'English Patient'

6.1 Introduction

In this chapter, a case study of England's recent experiences of a consultation process about the proposed designation of Marine Conservation Zones (MCZs) is analysed using the debate between NPs and SCs developed in Chap. 1 and the policy network analysis developed in Chap. 2. The English case was chosen for study for two reasons: partly because the waters around the UK (especially the North Sea) have been subjected to more intensive research scrutiny than any other inshore area around the world so data deficiency should be comparatively low; and partly because the UK demonstrates in a graphic way the controversy raised by proposals to introduce protected areas in developed countries with temperate climates where major commercial fisheries have been established for generations. From an analysis of the 'grey' literature and of key-informant interviews with 21 members of the English policy community who had input into the policy debates on MCZs, the contrasting NP and SC viewpoints are used as lenses to discuss contrasting opinions on five key themes regarding the planning of the MCZ network: objectives; data deficiencies; time-scales; consultation; and equity. The chapter builds on the work undertaken in Chap. 2 that explained the role that two policy network models—epistemic community (EpC) and advocacy coalition (AC)—had on influencing policymakers who work at the international level to write recommendations for MPA networks. It was suggested in Chap. 2 that the proposal to establish MPAs was initially due to the efforts of a committed group of marine experts (EpC), and that later the debate on MPAs extended to wider civil society as the green lobby became progressively more involved (pro-MPA AC), and provoked a reaction against indiscriminate MR advocacy by an anti-MPA AC. These two policy network models (EpC and AC) are applied to explain the actions of institutions that directly influenced English policy on MCZs. The findings of Chap. 6 suggest that policymakers failed to make clear the objectives of the MCZs in England; the impact of MCZs on local communities; and the distribution of the costs of MCZs. The consequent uncertainty generated hostility to the MCZ process in the fishing industry, which threatened its long-term success. In the event, the first tranche of decision-making saw only 27 of the 127

A. Caveen et al., *The Controversy over Marine Protected Areas*,
SpringerBriefs in Environmental Science, DOI 10.1007/978-3-319-10957-2_6

recommended MCZs being designated by the government, which suggested that the anti-MPA AC (the SCs) had largely won the first round, though further designations are expected during the next few years, depending on the robustness of the scientific evidence for them.

In the UK, landings per unit of fishing power of UK bottom trawl fisheries have been reduced by 94 % since the late 1800s (Thurstan et al. 2010), and industrial fishing has caused significant declines in the biodiversity of some of the UK's marine ecosystems (Thurstan and Roberts 2010) as well as the loss of large predatory fish (Jennings and Blanchard 2004). This general fall in the productivity of UK marine ecosystems has been coupled with the socioeconomic decline of many UK fishing communities—a social dimension often overlooked by decision-makers during the setting of marine policy objectives (Symes and Phillipson 2009). One proposed mode of dealing with the effects of fishing effort on fish stocks and marine biodiversity is through the establishment of networks of MPAs across the territorial waters of a country through a process known as 'systematic conservation planning' (Margules and Pressey 2000; Maiorano et al. 2009). The UK has been heavily criticised for falling behind other developed nations in its efforts to conserve marine nature and fish stocks by such planning (Roberts and Mason 2008; Wright 2010; Monbiot 2012). Jones (2008) reported that before 1981, there was no statutory provision for MPAs in the UK, and even when such statutory provision was made in 1981, only three MPAs were ever designated. Under the 1992 EU Habitats Directive, the UK has designated 80 multi-use MPAs, but there is only one NTMPA or marine reserve—Lundy, in the Bristol Channel, constituting 0.002 % of UK's territorial waters. However, following 10 years of international pressure (e.g. through OSPAR) and domestic pressure from the UK environmental lobby, a UK Marine and Coastal Access Act (MCAA) was granted royal assent in 2009 to help the UK achieve 'clean, healthy, safe, productive and biologically diverse oceans and seas' (Defra 2002, p. 5, para. 1.8). A key feature of the MCAA is that it placed a statutory duty on the UK government, and governments of the devolved administrations (Scotland, Wales, and Northern Ireland through their own respective marine acts) to establish networks of MPAs (or MCZs in England) (see Fig. 6.1) to protect a range of representative UK habitats. Greenpeace called for a network of MRs in the North Sea covering 40 % of its area. The Royal Commission on Environmental Pollution (RCEP) in 2004 called for 30 % of UK's 200-mile EEZ to be designated as MRs to form an ecologically coherent network for biodiversity and fish stock protection. The MCS called for 30 % of the UK's seas to be designated as MRs by 2020 (Jones 2009). By contrast, NE, while calling for a coherent network of MPAs, only specified MRs where necessary for monitoring and research purposes. These various pressures indicate the existence of both an EpC and a pro-MPA AC.

Before the proposed designation of MCZs, around 30 % of the territorial waters around England and Wales were already under some type of spatial management measures[1], including fisheries closures administered through the CFP (Rogers 1997), protected areas administered through the EU Birds and Habitats Directives

[1] Though critics pointed out that these 'spatial management measures' were not the same as MPAs (Lawton 2007), and they failed to protect UK marine biodiversity (Defra 2004).

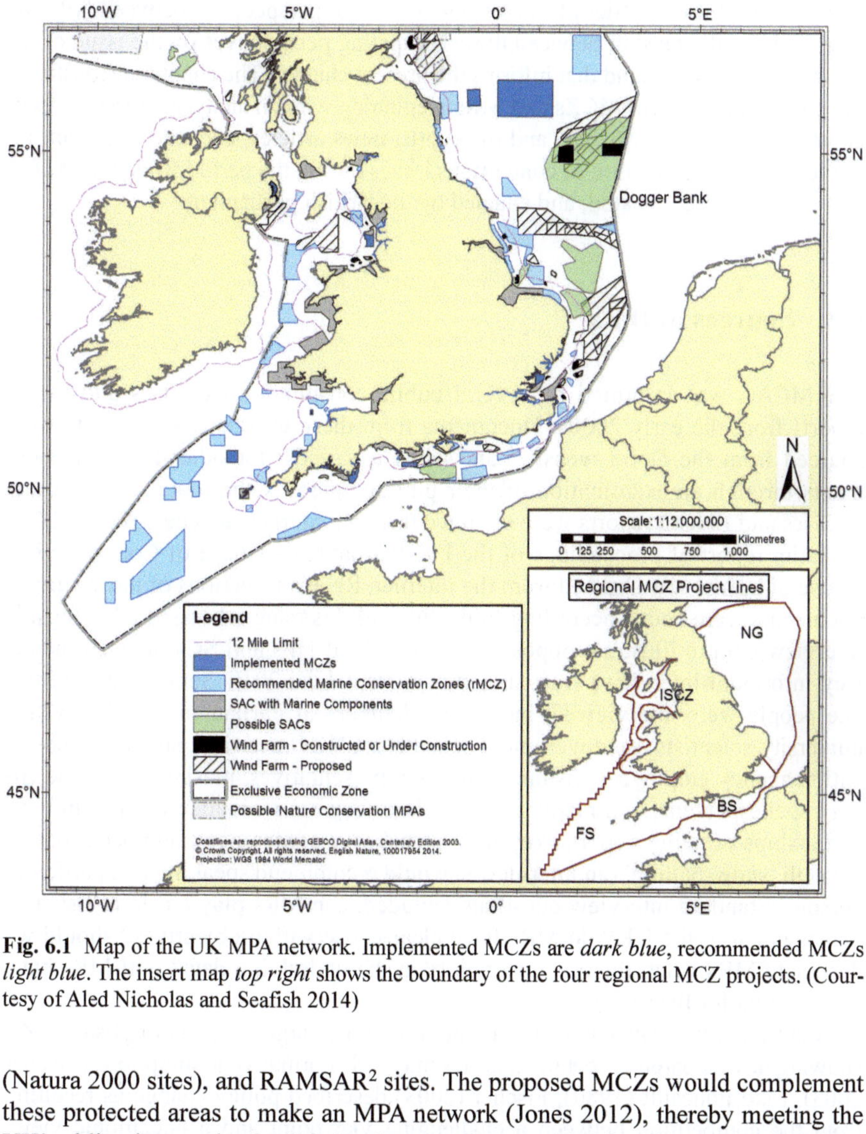

Fig. 6.1 Map of the UK MPA network. Implemented MCZs are *dark blue*, recommended MCZs *light blue*. The insert map *top right* shows the boundary of the four regional MCZ projects. (Courtesy of Aled Nicholas and Seafish 2014)

(Natura 2000 sites), and RAMSAR[2] sites. The proposed MCZs would complement these protected areas to make an MPA network (Jones 2012), thereby meeting the UK's obligations under the 'Marine Strategy Framework Directive' (MSFD) (2008), OSPAR (1992), the WSSD (2002), and the CBD (2010). However, although many ENGOs believed that MRs should take a more central role in UK marine management (Roberts and Hawkins 2000; Wright 2010), there was no explicit call for MRs in the MCAA. This is one reason why the MCCA was criticised by the environmental lobby for not providing adequate protection to UK habitats from damaging fishing activities and developers. The MCCA was also criticised by the fishing industry for having disproportionate negative socioeconomic impacts on fishermen (Jones

[2] The Convention of Wetlands of International Importance (1971).

2007). Behind this conflict of interest lay a clash of perspectives between NPs and SCs, and it is the tension between these competing perspectives on the issue of the MCZ network in England that informs the present chapter. The chapter deconstructs the policy debates over MCZs and MRs, beginning with an analysis of the antecedent documents to the MCCA, and of reports, news articles, and key informant interviews, to build a picture of contentious issues during the period when the MCCA was being drafted, debated, and enacted by the British parliament.

6.2 Sources of Data

The MCAA was informed by several public consultation exercises and expert reports from the early 2000s. Documents from these exercises and reports were sourced from the Defra website[3] and acted as 'seeds' for other relevant documents through an examination of their bibliographies. Other data such as press releases and ENGO reports were obtained through internet searches. Hansard (the verbatim report of proceedings of the UK Parliament's House of Commons and House of Lords) was sourced from the Internet. Relevant sections of these various documents were coded according to theme, and this material was used alongside interview data to illustrate opposing viewpoints of NPs and SCs on each theme. Key-informant interviews were undertaken from June 2010–Sept 2012. Twenty-one people were interviewed, including Members of Parliament, civil servants, university scientists, ex-government scientists, ENGO activists, statutory conservation agency employees, fishing industry representatives, marine renewable energy spokespersons, and media workers. Interviewees were identified from their authorships of policy reports, key-note speeches at conferences, and occasionally through 'snow-balling' (an interviewee would recommend speaking to a particular person). Standard interview questions included: can MRs play a role in fisheries management in the UK? do MRs have clear conservation objectives? should we designate MRs when there is uncertainty in the underlying data? will MRs have any benefits for fishermen?

Analysis of this data enabled us to interpret policy arguments on English MCZs, showing how apparently technical issues may hide normative assumptions (Fischer 2003). Like Foucault (1980), Fischer (2003) described policy change as resulting from the competition between a challenging viewpoint and a hegemonic viewpoint which is embedded in existing institutions. This builds on the idea by Hajer (1995) who suggests that politics is an argumentative struggle in which actors not only try to make their opponents see the problem according to their viewpoint, but also seek to portray other actors in specific ways: for instance, 'struggling, brave fishermen vs. unhelpful environmentalists' (Lawton 2007, p. 470), or fish robbers versus a well-intentioned European Commission (NFFO 2011a). From an analysis of relevant policy documents and grey literature, key-informant interviews, and

[3] http://archive.defra.gov.uk/environment/marine/legislation/mcaa/key-docs.htm

peer-reviewed science, an argumentative struggle between the two competing viewpoints of NP and SC emerged repeatedly. Both the NP and the SC discourses claim to have strong scientific foundations, but scientific claims that are based on robust empirical evidence can be differentiated from those that are less well grounded. We look at how different stakeholders viewed the term 'science'; the extent to which the debate became 'scientised' with respect to the planning of MCZs; and how advocacy caused empirical claims and normative beliefs to overlap.

6.3 Results and Discussion[4]

6.3.1 Planning Work Preceding MCAA Drafting (1999–2006)

Campaigns for a Marine Bill by the UK green lobby began in the late 1990s, and in 2000, an advocacy coalition, the Wildlife and Countryside Link, composed of organisations including MCS, RSPB, the Wildlife Trusts, the Whale and Dolphin Conservation Society (WDCS), and the WWF formed a Marine Task Force to coordinate the exercise. This network organised several public operations, including the Marine Reserves Now campaign run by the MCS and Co-op group (2007–2009), and Marinet's campaign calling for MRs. There was also a UK-based Marine Reserves Coalition (a partnership of six organisations: Blue Marine Foundation, Client Earth, Greenpeace UK, MCS, Pew Environmental Group, and Zoological Society London) campaigning for the government to establish an ecologically coherent network of Highly Protected Marine Conservation Zones (HPMCZs) covering at least 30 % of UK seas out to 200 nautical miles. In 2009, this pro-MR lobby sought to persuade the public to urge their MPs to sign an Early Day Motion (EDM 337[5]) registering support for an amendment to the Marine Bill to include this requirement. In an email to a prospective supporter, a campaigner from Marinet wrote that 'We'll be in touch again with the details of what to ask your MP. As well as electronic campaign materials available on the website, we have generous amounts of printed resources that can be sent out to you' (pers. comm. 11.2.09).

At the same time, there were growing calls for MRs made by international regimes such as OSPAR, CBD and WSSD epitomised by a WWF report that synthesised the current evidence (Roberts and Hawkins 2000). Responding to these

[4] Quotations from interviewee transcripts are italicised.

[5] EDM 337: "That this House notes the recommendation made by the Royal Commission in their 25th Report in 2004 that a widespread network of highly protected marine reserves throughout all UK seas is an important management tool which is required in order to rebuild commercial fish stocks and to halt the serious damage being caused to marine ecosystems; and calls upon the government to develop selection criteria under the Marine Bill for establishing a network of protected marine areas around science-based decision making" (Marinet 2009).

Table 6.1 Principles of UK marine strategy relevant to the science–policy interface

Principle	Description	Conflicts
Robust science	Understanding the processes and influences that impact on the marine environment and using research to inform policymaking and marine management	Science is open to varying degrees of interpretation, particularly regarding the debates on the efficacy of MRs and sensitivity of different habitats to different fishing pressures
Precautionary principle	Sensibly erring on the side of caution where the scientific evidence is not conclusive	Members of the fishing industry argue that it is unrealistic and unfair to shift the burden of proof on to fishermen: how do you prove a gear has no impact?
Stakeholder involvement	Involving all stakeholders so that they are an integral part of the decision-making process	Does inclusive decision-making diminish the role of the expert? Who is a stakeholder?

pressures, during the late 1990s Defra began laying the foundations for the MCAA by commissioning a series of reports that set out the UK's vision and strategy for improved marine environmental management. The first of these reports, *Safeguarding our Seas,* stated that the UK government was committed to an ecosystem-based approach (EBA) to marine management, a key element of which 'is the conservation and where possible, enhancement of marine ecosystems in a way that conserves biological diversity and ensures the sustainable development of our marine resources' (Defra 2002, p. 9, para. 1.32). This document stated six principles that would inform the government's approach to EBA, three of which are directly relevant to the science–policy interface (see Table 6.1).

Defra pledged its commitment at the 5th North Sea Conference held in Bergen in 2002 to identify and designate MPAs by 2010 according to a 'clear understanding of natural processes and the ecological requirements of marine species, habitats and ecosystems' (Defra 2002, p. 9, para. 1.33). In 2003, OSPAR adopted MPAs as an approach to protect marine biodiversity in the NE Atlantic: 'the Commission will, *inter alia*, promote the establishment of a network of MPAs to ensure the sustainable use, protection, and conservation of marine biological diversity and ecosystems' (OSPAR 2003a, p. 1, para. 2). Such a network would be ecologically coherent and would restore and prevent further degradation of species and habitats. The UK government's commitment to OSPAR was the main driver behind the push by the devolved administrations to establish an MPA network (Defra civil servant), and it was the job of the Joint Nature Conservation Committee (JNCC) to advise government on the designation of the UK-wide network of MPAs[6].

In late 1999, the UK Government and devolved administrations commissioned a working group of the statutory nature conservation agencies (SNCAs) and commercial and recreational interests to conduct a pilot study in the Irish Sea to develop a framework for planning that would reconcile nature conservation objectives with

[6] The four devolved administrations of England, Wales, Scotland, and Northern Ireland have each used different approaches in setting up their respective MPA networks.

Table 6.2 Principles of an ecologically coherent network of important areas. (Defra 2004, adapted from Ballantine 1999)

Design principle	Description
Connectivity	Networks should be designed to ensure that areas are mutually supporting (i.e. populations of animals and plants in one area should be capable of supporting, and be supported by populations in other areas). The need to protect vulnerable life stages of highly mobile species, including their movement between breeding and feeding grounds, should be taken fully into account
Representation	Networks should seek to incorporate the full spectrum of biological diversity, not just that subset which relates to rarity, endangerment, or other pre-selected importance values
Replication	Examples of habitats (or concentrations of species) should be replicated in separate areas to insure against loss due to catastrophic events whether from natural or human-induced causes
Sufficiency	The total area of the network, and its distribution in terms of individual component areas, should be capable of meeting the objective of sustaining species and their habitats in perpetuity
Practicality	The best available information should be used in site selection, but the development of the network should not be delayed pending action to collect further information, though practical considerations, including those which support sustainable development, should also be taken into account in site selection

development objectives in the marine environment (Vincent et al. 2004). The subsequent report, *Review of Marine Nature Conservation* recommended the establishment of an ecologically coherent network of marine protected areas designed according to the following five principles (see Table 6.2): connectivity, representation, replication, sufficiency, and practicality, which were taken from experiences of designating MRs in New Zealand (Ballantine 1999). These principles formed the basis of England's approach to setting up MCZs: in March 2010, Defra issued a list of seven principles which would underpin the design of the MPA network—representativity, replication, viability, adequacy, connectivity, protection, and use of best available evidence (see Jones 2012)—and became the foundation of the Ecological Network Guidance (ENG) which was to inform the deliberations of the four regional projects: Finding Sanctuary (South-West); Irish Sea Conservation Zones (Irish Sea); Net Gain (North Sea); and Balanced Seas (Eastern Channel).

However, in the ENG, there were no recommendations for NTMPAs, as the government wanted to retain flexibility to decide the appropriate kind of protection on a case-by-case basis. In a letter to the Friends of the Earth Marine Network (Marinet), the Minister (Huw Irranca-Davies) stated that 'we are committed to ensuring that the Bill provides the right level of protection for MCZs, according to the conservation objectives of each zone and the science underpinning it. I have no doubt that there will be MCZs with high levels of protection' (Irranca-Davies 2009). The only mention of MRs in the *Review* was a call for the establishment of large-scale trial MRs to test their fisheries benefits, to which the government responded by requiring at least one 'reference area' (i.e. a no-entry zone) to be established in each regional project area as a bench mark against which to judge the effectiveness of partially protected MCZs. The 127 proposed MCZs contained 65 such reference areas within them, 2% of their total areas—well short of the 30% demanded by RCEP (Jones 2012).

MRs were much more emphatically on the agenda of the ENGOs and the statutory conservation agency, English Nature (now Natural England (NE)) (Rees et al. 2013). Many ENGOs had fought for the inclusion of NTMPAs in 30% of English waters (Jones 2012), and English Nature argued that only MRs could deliver the EBA (Laffoley et al. 2004). Complementing this positive narrative about the benefits of MRs, a new negative narrative emerged in the UK in 2004. There was a growing appetite in the national UK media for stories that reflected disparagingly on activities of the fishing industry[7] (Symes 2005). In the words of a journalist respondent:

> *the media, especially the national...media and the TV, are very much on the environmental bandwagon. They'll always tend to be interested in stories that show fishermen in a bad light...guys like Charles Clover, at The Telegraph, and from The Times, Frank Pope, they're very much wedded to the environment and very anti-fishermen. I think that is a significant change that's happened over the last 10 or 20 years, the public's view of fishermen. At one time they were pretty well respected: they were guys who went to sea and did a tough job and put food on the table; horny-handed heroes, salty sons of toil. But I'm afraid now, many people just think of them as brigands and vandals, destroyers of the earth, sadly. And I think the media's had quite a large part to play in promoting that image of fishermen.*

This negative view of the fishing industry was also reflected in two influential reports published in 2004 which examined the impact of fisheries on UK marine ecosystems. First, the *Net Benefits* (2004) report from the Prime Minister's Strategy Unit (PMSU) stated that 'commercial fishing has had the largest single negative impact on the marine environment's sustainability' (PMSU 2004, p. 10, para. 5). It listed 33 recommendations that would facilitate the reform of the UK fishing industry, one of which was for a trial run of MPAs. It is unclear in this report precisely what a MPA meant, though the context seems to suggest an area where there is at least a ban on towed bottom gear. Acknowledging the uncertainties of using MPAs in UK waters, the report asserted that 'as with other aspects of fisheries management, a lack of perfect knowledge should not lead to inaction and maintenance of the status quo, but to an adaptive and precautionary approach' (PMSU 2004, p. 93, para. 6).

Second, the scientifically authoritative RCEP report (2004, p. 205, para. 8.96), *Turning the Tide,* called for 30% of UK waters to be set aside as MRs in order to combat overfishing, and claimed that the recommendations of the *Net Benefits* report were 'too tentative and too slow'. However, the RCEP report was criticised for overlooking the differences between reef-based areas and soft bottom areas (Defra 2010); for making scientific claims that were not supported by peer-reviewed sources; for referencing unpublished work which could not be found (Roberts and Mason 2008); and for lack of transparency in its calculations of the costs of implementing

[7] For example, the headlines in *The Daily Telegraph* and *The Sunday Times* reported that there were only 100 large cod left in the North Sea (Leake 2012).

and managing the marine reserve network (Symes 2005). Nevertheless, the RCEP report provided prestigious support for a narrative in which fish stocks were crashing and MRs were needed to deliver stock recovery and ensure that fish populations were sustainable in the long term (Lawton 2007). This narrative was adopted by members of the green lobby in their attempts to persuade the UK government to establish networks of MRs, and it formed a significant part of the NP viewpoint.

Responding to *Turning the Tide,* the UK government stated that it was 'developing plans for a controlled trial of MPAs which aim to have both fisheries and wider marine conservation benefits' (Defra 2006c, p. 9, para. 8.96). However, this controlled trial was never carried out—a failure blamed on the uncooperative attitude of the fishing industry, according to an NP scientist respondent: '*So what would the effect be of putting large no-take zones?... We made the recommendation that to resolve this issue once and for all the fishing industry should engage with the conservation sector to do some large-scale no-take trials to see what the benefits were—they haven't done it*'. There were, however, financial reasons why these trials for MRs did not go ahead (MacGarvin and Jones 2000); if such a large-scale trial were to be undertaken there would have to be some reassurance from government that fishers' loss of profit would be fully compensated, but the issue of compensation raised several objections. First, UK marine fisheries are common property, so those exercising their right to use the resource are not eligible for compensation from the public purse. Second, fishers could switch to alternative grounds to maintain their income. Third, it is very difficult to distinguish between fishermen who have a legitimate right to compensation and those who lie about their fishing in a certain area in order to receive compensation (Jones 2009). Fourth, fisheries management in the UK is under the exclusive legislative jurisdiction of the European Commission through the CFP[8], so if MRs were to be implemented in the UK EEZ, compensation for the negative impact on other countries' fishing fleets' loss of income would also be due (Symes 2005). Fifth, the fishing industry had already benefitted from subsidies: '*From a societal perspective, the tax payer has already paid for subsidies to keep the towed sector of the fishing fleet in profit. Overexploitation by the fishing industry has left us in the position we are currently facing, therefore they should not be compensated*' (ENGO policy officer).

The case for MRs also ran into opposition over the proposed percentages of sea area for MRs. Gell and Roberts (2003) suggested from their review of modelling studies that between 20 and 50% of the ocean should be designated as MRs. The case in favour of a large percentage was rehearsed by a NP marine scientist respondent:

> *if you look at it from a range of scientific angles, whether it is on the grounds of increasing the spawning stock biomass of target fish species, maximising long term yields, minimising*

[8] The CFP can make provision for area closures for the protection of nature under its emergency measures; that is, "if there is evidence of a serious threat to the conservation of living aquatic resources, or to the marine ecosystem resulting from fishing activities and requiring immediate action" (EC 2002, p. 1, para 10).

loss of genetic heterogeneity, ensuring that you have all species protected somewhere and replicates of them in different protected areas...the answers that you get on how much you need are in the tens of percent not just percent...20, 30, 40; it's not 2, 3 or 4.

However, this claim does not take into account such factors as the patchy distribution of fishing effort (Jennings and Lee 2012); the complexity of the fishing industry (Phillipson 2002); and habitat heterogeneity (Greenstreet et al. 2009). The trawling footprint has been estimated at between 5.4 and 21.4 % for English and Welsh waters (Eastwood et al. 2007; Stelzenmuller et al. 2008; Jennings and Lee 2012). Although this is likely to be an underestimate for two reasons: first, the estimate excluded static and pelagic gears; and second, only vessel monitoring system (VMS) data (not required for vessels < 15 m) was used, thus ignoring the distribution of effort by smaller trawlers (Jones 2008), it means that large portions of seabed were not trawled regularly, if at all, and that de facto MRs already existed in English waters. Advocacy for MRs to reduce fishing pressure also ignored the fact that since 1996, the UK fishing fleet had contracted by 26 %, and fishing power had fallen by 23 %, with a parallel reduction in pressure on the marine environment (Elliott et al. 2012).

Defra's response to the RCEP recommendation for designating 30 % of the UK's EEZ as MRs (Defra 2006c) was cautious, stating that it was uncertain of the scientific basis for that figure. Perhaps Defra's caution was based on the findings of three studies it had commissioned to review the lessons learned from MRs in Northern European waters, and to calculate the potential contribution that MPAs could make to the recovery of specific fish stocks (CEFAS 2005; Pascoe and Mardle 2005; Sweeting and Polunin 2005), since the general theme that emerged from these reports was that candidate reference areas should be assessed for their merits on a case-by-case basis.

Clearly, a challenging voice had emerged to contest the dominant[9] NP discourse in the pre-drafting stage of the MCAA: the fishing industry and other resource users set out to influence government by forming an anti-MPA advocacy coalition (AC)[10]. Initially, the fishing industry collectively was slow to recognise the potential impacts of MCZs on their interests, perhaps because of limited communication between Defra and the fishing industry as the Marine Bill was being planned. Later, however, the fishing industry made up for lost ground by challenging the scientific and democratic basis of the MCZ network designation process through the MPA Fishing Coalition (MPAC). The MPAC was officially launched at the House of Commons in February 2010, 'to provide the fishing industry with a powerful negotiating platform' (*Fishing News* 26.2.10: 1). However, it was conceived not only as a pressure group but also as a forensic body with scientific expertise to expose scientific error in the MCZ designation process: 'the only way fishermen are going to avoid being swept off fishing grounds is by presenting a strong, coherent

[9] 'Dominant' in the sense that the nature protectionist discourse coalition has been successful in influencing the UK government to establish networks of MPAs. However, as we shall see, NPs argue that they have not had their own way—particularly in determining how many MCZs should be designated and in deciding what management measures should be implemented in MCZs.

[10] Weible and Sabatier (2005, p. 193) reported a similar configuration of ACs in California: "We find *at least* two advocacy coalitions in the California MLPA policy subsystem…an anti-MPA advocacy coalition and a pro-MPA advocacy coalition" (italics in original).

case backed by factual evidence…It is clear that just lodging a protest—even a very loud protest—is simply not going to be enough. The MPA Fishing Coalition will…not deal in loud diplomacy. It will be serious and evidence-based' (*Fishing News* 26.2.10: 1). Importantly, the MPAC did not oppose MPAs per se (*Fishing News* 10.12.10: 24)—they 'are a reality' (*Fishing News* 2.4.10: 24)—but insisted that 'MPAs must be introduced in a fair, considered and proportionate way' (*Fishing News* 4.12.09: 6). Austin Mitchell (MP for the fishing port of Grimsby) said that 'The fishing industry supports Marine Conservation Zones (MCZs) but not if they're to be used as environmental overkill' (*Fishing News* 6.11.09: 6). Lockwood (MPAC chairman) stated that 'We are not here to thwart MPAs, but to ensure they are science-led by genuine evidence, and that whatever restrictions are introduced are rational and proportionate' (*Fishing News* 25.11.11: 8). He said that 'The industry is not against MPAs in principle but it is totally opposed to designation by diktat…I have found that too many people engaged in MPA promotion and site selection have virtually no knowledge or understanding of fish, fishing or the fishing industry…the MPA process invariably starts with a blank sheet of paper rather than building on what is already in place…Skippers have knowledge and data that must form a vital part of the MPA designation process' (*Fishnewseu.com* 15.6.10). The MPAC did not intend to precipitate an 'adversarial trial of strength' but to adopt 'a consensus-based approach with negotiated and agreed outcomes' (NFFO 2012a).

The pro-MPA AC role adopted by some scientists was criticised by the MPAC because it sought to increase fish stocks through reducing fishing mortality by siting MCZs in areas where fishing mortality was the highest, but the fishing industry pointed out that aside from the negative socioeconomic impact of designating an MCZ in an area that is intensively fished (the NFFO declared that 'We are determined that this will not be some kind of marine highland clearance, with fishermen callously evicted from their traditional fishing grounds' (*Fishing News* 26.2.10: 2)), the displacement of fishing activity to areas outside the MCZ could lead to the loss of biodiversity in habitats that were previously un-fished (NFFO 2010). The anti-MPA AC also criticised the pro-MPA AC for its shift in advocacy on the issue of trawling and dredging. Initially, the majority of NP scientists wanted a blanket ban on trawling and dredging in all MCZs because in their view, if this did not happen, MCZs would merely be 'paper parks' (Monbiot 2012). However, scientific studies showing the impact of bottom trawling on the seabed produced mixed results—some suggesting that trawling had a negative impact on habitat structure (Watling and Norse 1998; Hall-Spencer and Moore 2000) and benthic invertebrate productivity (Hiddink et al. 2006), but others suggesting that any impact was negligible (Hiddink et al. 2007), or even positive (Beare et al. 2014). On the realisation that in many cases no-trawl MCZs could not be justified on the evidence, NP scientists reframed the debate calling for no-take reference areas to be established to allow the recovery of the seabed for research purposes.[11] But the fishing industry adamantly rejected this argument:

---

[11] For example, the South Orkney MPA in the Southern Ocean was justified as "a special opportunity to study the effects of climate change free from the influences of other forms of human activity" (WWF 2010, p. 26).

> There is absolutely no need for draconian areas of 'no-man's land' for the purpose of measuring ecological improvement in marine protected areas which would be readily evident from any general time-series programme of monitoring. Such proposals…show a careless disregard for people's livelihoods in order simply to give a free reign to conservation scientists' experiments…[Is this] really what society wants from marine protected areas; scientific playgrounds and barren fishing communities that were once vibrant? (Rodmell 2011)

The MPAC was strongly opposed to reference areas: 'They are not based on credible science directly applicable to the UK…Whether they are effective or not should be tested against evidence from temperate zone North Atlantic, not tropical reefs, but there is no sign of this. There is also no requirement for reference areas within the Marine Act' (*Fishing News* 25.11.11: 8). MPAC said that 'The fishing industry can be persuaded to accept restrictions on its activities where a specific scientific case has been presented, argued and found substantive. What it cannot accept is the imposition of restrictions on a whim—as appears to be the case with MCZ 'reference areas'' (MPAC n.d.). Rodmell described reference areas as "a no man's land' banning the vast majority of activities for the sake of establishing a set of scientific research dominions' (*Fishing News* 13.1.12: 17). JNCC and NE rejected this attack on reference areas: 'Reference areas are not scientific playgrounds. They will help ensure that decisions about the way MCZs are managed are based on the best available evidence' (*Fishing News* 18.3.11: 24).

Another criticism of the pro-MPA AC was for its overly conservative attitude to risk. For example, several MCZs were proposed by scientists on the basis of limited evidence of the presence of a habitat/species vulnerable to fishing[12]. The scientists justified their stance on grounds that imperfect knowledge was reason enough to invoke the precautionary principle so that restrictions were placed on potentially damaging fishing methods (e.g. dredging/trawling). However, the fishing industry claimed that the absence of robust evidence in support of placing a restriction on fishing was reason enough to allow current fishing to continue (see Chap. 6). On this issue, neither side could legitimately claim the backing of science, because the dispute was over competing (subjective) attitudes to risk, not competing claims of (objective) scientific evidence.

Figure 6.2 gives a graphic indication of how the two ACs influenced outcomes on MCZs by the relative size of the dashed boxes and positions. Currently, there is a stronger pro-NP than pro-SC narrative running through the national media, though there is still uncertainty over the respective influence of the pro-MPA and anti-MPA ACs on government, since what types of fishing activity will be excluded from MCZs have not yet been decided. It is predicted that the final management measures are likely to reflect a compromise between the two factions.

Scientists could be found offering their expertise to both the pro-MPA and the anti-MPA ACs, thereby bolstering the credibility of the cause of each coalition and influencing public opinion (Grorud-Colvert et al. 2010). Both ACs played up certain kinds of evidence to support their respective causes: the pro-MPA scientists

[12] Indeed, according to the industry, some MCZs missed their intended conservation feature altogether (NFFO 2012a).

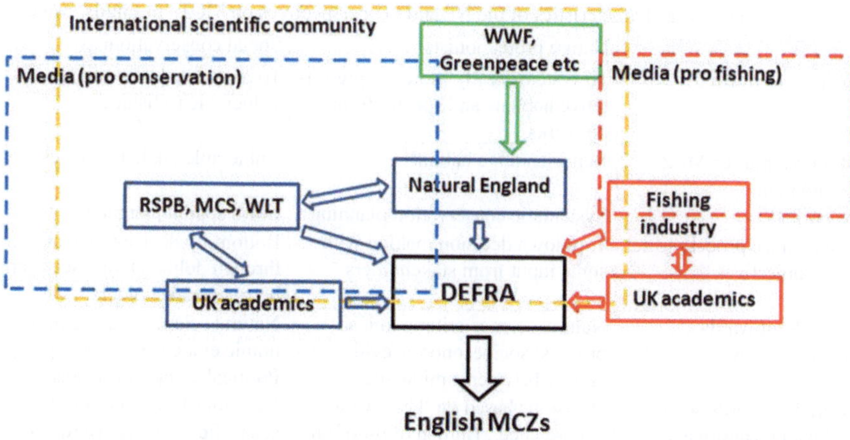

Fig. 6.2 The advocacy coalitions that have shaped outcomes on the design and management of MCZs. The *green* box indicates international ENGOs that have influenced thinking on MPAs both internationally and nationally. *Blue* boxes indicate the pro-MPA advocacy coalition; *red* boxes indicate the anti-MPA advocacy coalition

emphasised the fisheries benefits of marine reserves; while the anti-MPA lobby emphasised the displacement effects of marine reserves (Turnhout et al. 2007; Weible 2007).

So two discourses have emerged from this discussion: (1) the dominant NP discourse, whose main adherents were environmentalists and marine ecologist scientists, which called for the systematic protection of representative habitats at a national level and emphasised the use of ecological criteria to lead site designation; and (2) the challenging SC discourse, whose main adherents were marine users (principally the fishing industry) and fisheries scientists, which was not necessarily anti-MR, but keen to prevent inappropriate siting of MRs by emphasising that socioeconomic and other evidence (for example, habitat, species distributions, and climate change effects) must be taken into account during site designation, and that such planning was best done at the local level (see Table 6.3).

The respective influence of the NP and SC perspectives on the planning of MCZs is now discussed.

6.3.2 *Planning of MCZs (2006–present)*

The process of selecting sites for possible MCZ designations took the form of four non-statutory regional projects in England ('Net Gain'—the North Sea; 'Irish Sea Conservation Zones'—the Irish Sea; 'Balanced Seas'—the English Channel; and 'Finding Sanctuary'—the South West), each of which involved three groups: a regional stakeholder group; a regional project team; and a regional project board. The stakeholder group worked on MCZ site-selection, supported by the project team

Table 6.3 Contrasting characteristics of the NP and SC perspectives on MCZs in English waters

	Nature protectionists	Social conservationists
Objectives of MCZs	To systematically protect representative habitats and species through networks	To protect habitats and species vulnerable to fishing
Main criteria for MCZ designation	Representative habitats	Vulnerable habitats/species
Approach	Systematic conservation planning	Local spatial planning
Governance process to set MCZ objectives	Top-down decision making with some input from stakeholders	Bottom-up decision making through deliberative discussion between stakeholders
Attitudes towards science-policy	Natural science criteria to lead process; socioeconomic evidence to choose between similar sites	Natural science and socioeconomic evidence treated equally. Political compromise necessary
Attitude towards science and the precautionary approach	Decisions based on 'best available science'. Burden of proof on the fishing industry to show that activities don't cause damage to a conservation feature	Decisions based on robust scientific evidence. Burden of proof on protectionists to show that a feature sensitive to fishing exists
Attitude towards conservation	Ecosystem preservation necessary for sustainable use	Some impact inevitable, though should protect sensitive habitats
Scale	National/regional	Local
Time frame for decision	Relatively short	Long
Narratives from scientific literature	Spillover benefits, ecological coherence, habitat destruction	Displacement, effect on local communities, wider economic impacts (e.g. food supply, and users moving elsewhere)
Criticisms from opposing discourse	Preservationist, inhumane, ignores the needs of local people	Favours short-term economic interests, potentially could miss strategic conservation goals

and the project board's scientific expertise. The proposals emerging from the four regional projects were to be assessed by the independent Science Advisory Panel (SAP) (Jones 2012) (whose members included experts on MRs) for their consistency with the ENG (prepared by NE and JNCC), after which SAP would publish its own recommendations to Defra, along with socioeconomic impact assessments (Seafish 2009), for its MCZ designations after a further process of public consultation. It is worth noting that the ENG was derived from a considerable international literature based on planners/managers/scientists' experiences of developing networks of MPAs in California (Airame et al. 2003) and Australia (Fernandes et al. 2005), and focused on how such networks could best be designed according to ecological (Airame et al. 2003) as well as socioeconomic criteria (Lundquist and Granek 2005). Several papers from the Working Group on Marine Reserves published in a special issue of *Ecological Applications* (2003) were used to inform the ENG that provided the principles for the design of the English MCZ network, and at least two scientists from the USA—S. Gaines and M. Carr—visited England to disseminate their experience of being involved in the planning of the Californian

MPA network. Indeed, the English MCZ planning process was largely modelled on that used in California[13].

In this period (2006–present), we focus on five controversies between the NP and SC perspectives over the planning of MCZs: the objectives of MCZs; data deficiencies; time-scales; consultation; and equity. Running through these controversies we can see both empirical and normative disputes.

Controversy over the Objectives of MCZs

In 2005, with cross-party support, the Labour Government made a general election manifesto commitment to a Marine Act. MPAs were a central part of the proposed Marine Bill, and the first consultation in 2006 showed that the government was intent on establishing a network of MPAs to promote the recovery of vulnerable species and habitats, representative species and habitats, physical marine features and ecological processes, and the protection of spawning and nursery areas (Defra 2006b). Pieraccini (2013, p. 106) affirmed that this was a serious ecological commitment: 'the MCAA, by providing for the establishment of new nationally important MPAs with a broad ecological focus, has moved beyond a minimalist approach to the establishment of an ecological coherent MPAs network' (see also Mangi et al. 2011; McVittie and Moran 2010).

However, ambiguity over the objectives of MPAs was present from the beginning. One confusion was over the status of the MCZs. As Lockwood (2013c) pointed out, some NPs (such as Charles Clover, George Monbiot and Hugh Fearnley-Whittingstall) gave the impression that 'we have no marine protected areas (MPAs) in UK waters and that the proposed list of 127 MCZs for English waters are a novel concept without which our marine environment is doomed to apocalyptic destruction…this notion does not stand up to *factual scrutiny*' (italics in original). The fact is that the MCZs were designed to supplement the 131 marine sites of special scientific interest (SSSIs); the 46 special areas of conservation (SACs); and the 43 special protection areas (SPAs) already in place in the UK, which make up 20% of English coastal waters. There are also many other spatial restrictions on fishing imposed by authorities who regulate fisheries, energy installations, defence activities, harbours, shipping, navigation, and communication cabling—'restrictions that enable these sites also to make a contribution to marine biodiversity and conservation' (Lockwood 2013c).

A second confusion over MCZ objectives was that official documents appeared to affirm that the sole objective of MPAs was ecological—'The UK and Welsh Governments are committed to establishing an ecologically coherent network of effectively managed Marine Protected Areas (MPAs) by 2012. This network will…conserve sufficient rare, threatened, and representative species and habitats to maintain and improve biodiversity and ecosystems' (Defra 2008b, para. 2.2).

[13] For a detailed discussion of the California process for setting up an MPA network see Weible (2007, 2008; Weible and Sabatier 2005; Weible et al. 2004).

Yet the same documents also refer to socioeconomic objectives: 'It will generally be desirable to avoid designating MCZs with conservation objectives that are (significantly) incompatible with ongoing or future anticipated socioeconomic activities' (Defra 2008b, para. 5.20). JNCC (2010, p. 2) stated that there should be 'fair treatment of the range of socioeconomic interests throughout the planning process'. This controversy fuelled a quarrel over the insertion of references to socioeconomic considerations in the bill. SCs were particularly concerned with the wording of clause 117(7) of the Marine Bill, which read: 'In considering whether it is desirable to designate an area as an MCZ, the appropriate authority *may* have regard to any economic or social consequences of doing so' (Defra 2009, italics in original). SCs wanted 'may' to be changed to 'must', in order to place a statutory requirement on the government to take into account socioeconomic activities during the designation on MCZs. But NPs lobbied government to remove the clause altogether: '*We were asking for this clause to be removed on the basis that sites should be selected purely on science; and that it* [the clause] *would create a weaker site selection process than currently used for European sites…we were basically looking for some clarity that science and biodiversity needs would be the pure designating factor for the sites rather than socioeconomics'* (ENGO manager). In the event, the clause remained unchanged in the MCAA (Defra 2009), which Pieraccini (2013) saw as an important departure from conventional nature conservation law. This was a deliberate part of government thinking, as a letter to the NFFO (9.2.09) by a senior official at Defra indicated: 'our objective is to ensure that we minimise the economic and social impacts of MCZ designation' (Rodmell 2009). Defra's perception was that it aimed to achieve *both* ecological *and* socioeconomic objectives, as its definition of a MPA indicated: 'A clearly defined geographical space, recognised, dedicated and managed…to achieve the long-term conservation of nature with associated ecosystem services and cultural values' (Seafish 2009). In the summary of responses to the consultation process, stakeholders endorsed this strategy, stating that 'the UK Government should be seeking win-win situations with benefits to all three pillars of sustainable development (economic, social and environmental) or, at the very least, should be striking a reasonable balance between them' (Defra 2006a, p. 6, para. 2.9).

However, there remained a fundamental divide between NPs and SCs over what that balance should be, together with an underlying anxiety that the other side's interpretation would prevail (Defra 2006a). For example, from the NP perspective, Appleby and Jones (2012, p. 73) complained that 'the ecosystem approach has progressively been dropped in the drafting process of the legislation', but from the SC perspective, NFFO complained that NE/JNCC and Defra had already made up their minds in favour of an extensive network of MRs at the expense of the fishing industry. James Marsden, NE's Marine Director, sought to reassure the SC lobby by stating that 'it is not NE's intention that every conservation site should be a no-take zone either in part or in whole. It is possible that some sites will have no restrictions on fishing at all, while others will limit the use of some gears but not others' (*Fishing News* 12.2.10: 5).

A third controversy over objectives arose because of NPs' assurances that MCZs would benefit fisheries. A Defra-commissioned study to provide the evidence base for the Marine Bill excluded commercial fisheries from its investigations: 'Commercially important fish species on the OSPAR list were…excluded from the exercise because management of these stocks primarily represents a fisheries management objective rather than marine nature conservation' (Richardson et al. 2006a, p. 9). Yet the draft Marine Bill consultation proposed to 'establish a mechanism for MPA designation that could in addition potentially incorporate objectives such as the protection of commercially important fish stocks' (Richardson et al. 2006a, p. 14). Moreover, the regulatory impact assessment undertaken for the Marine Bill stated that there would be net fisheries benefits from MCZs, and although it conceded that the extent of off-site benefits (i.e. through spillover) was highly uncertain, it still provided an estimated net benefit of £ 16.8 million per year (Defra 2008a). However, it is difficult to know what such a figure meant, since it depended on where MCZs were designated and what management measures were put in place: even at the local level, economic assessments were particularly difficult to undertake, and results were widely disputed[14] (Fleming and Jones 2012). The NFFO took exception to exaggerated claims of fisheries benefits made during the planning of the Marine Bill: 'the claims made about their capacity to rebuild commercial fish stocks are hugely overblown and mainly depend on evidence from tropical reef fisheries whose relevance to most of the species is slight' (NFFO 2009b). However, the heavy promotion by NPs of MCZs for their expected fisheries benefits paid off in parliament debates, as shown by an extract from a speech by a member of the House of Lords, who revealed that he had read the book *Unnatural History of the Sea* (Roberts 2007b) which advocated MRs for their fisheries management benefits: 'the primary purpose of this legislation is to ensure the conservation of our fish stocks so that they can develop and rebuild after centuries of depredation by man…the fish stock that is built up successfully within the marine reserve area will spread out beyond that and provide happy hunting ground for fishers' (Lord Eden, member of the House of Lords on the UK's Marine Bill, Committee stage, 6th day; Eden 2009).

A fourth issue over objectives involved the inclusion of the term 'marine reserves' (MRs) in the Act. NPs called for additional wording to be added to clauses 116 and 117 of the Bill to include highly protected MCZs (Marinet 2009, p. 2). However, (as we saw earlier) despite considerable pressure, the government did not accede to these NP demands, but maintained its flexibility stance that 'the management measures needed for MPAs may vary widely depending on the objectives of each site and the sensitivity of the protected features to different activities and levels of disturbance. In some cases this may mean that only seasonal or time limited restrictions are required, if any. In others it could lead to complete restriction of activities on a site, which is commonly referred to as a highly protected marine reserve' (Defra 2006b, p. 101, para. 10.65).

[14] For example, the economic impact of the MPA proposals at Lyme Bay, Devon was estimated by four independent reports, with the estimates differing by an order of magnitude (Rees et al. 2010).

A fifth issue with MCZ objectives was that one of the two main elements of the ecological objective—'coherence'—was problematic. OSPAR called for an ecologically coherent network of MPAs to be established by 2010; NE called for a coherent network of MPAs by 2012; and the EU's MSFD required Member States to set up coherent MPA networks by 2020. Since the UK was committed to the OSPAR and MSFD targets, it was supposed to design MPA networks that met the coherence criterion. But the coherence requirement was confusing (Jones and Carpenter 2009). What exactly did the term 'coherent' mean? Did it simply mean 'interlinked', or did it mean *sustainably* interlinked, and if so, how was that sustainable linkage to be defined? Ardron (2008a, pp. 47–48) pointed out that 'though 'coherent', 'coherence' (and one instance of 'ecological coherence'...) is used throughout the EU Habitats (1992) and EU Birds (1979) Directives, these terms are not explicitly defined...The term does show up in the grey literature reports often in the context of Natura 2000, but is also not clearly defined' (see also Ardron (2008b). Ardron (2008a, p. 48, 51) said that 'Thus, achieving the goal of ecological coherence is one that ultimately cannot be measured exactly, but must rather be approached in a stepwise fashion, stated as a converging likelihood, based on a growing suite of approaches and indicators'. Ardron (2008b, p. 1527) also observed that because of data constraints, OSPAR's tests for ecological coherence would have to be more 'heuristic' than objective, 'based at least in part on subjective experience'. When a marine scientist was asked to explain the meaning of the term, 'ecological coherence', he was remarkably candid: '*Nobody really knows* [laughs]... *even though nobody actually knows what that is...I think what it means is a self-sustaining network of sites that are going to act as long term refuges of the species involved'.* Some scholars have questioned whether ecological coherence can be achieved because of gaps in our knowledge regarding the distribution and movement of species found in UK waters at different stages of their life-cycle (Jones and Carpenter 2009).

Such doubts confirmed the NFFO's suspicions of the ENG and the scientists behind it: The certainty that this vision will deliver an ecologically coherent network apparently no-one is to question, nor give consideration to other important needs such as sustainable fisheries. It is about time that those hiding behind this "science is right" charade began to recognise that humans do form part of the marine ecosystem' (*Fishing News* 13.1.12: 17). Seafish criticised the ENG for lack of transparency in the way it had been reviewed: 'First, the overwhelming majority of the organisations from which comments were invited are professional conservation institutions or groups, with little or no inclusion of representatives from fisheries organisations. Second, "the three key research reports which have been used to inform the guidelines within the ENG" have not been published. Third, the authors/co-authors of these three reports include at least one scientist (C.M. Roberts) who is a well-known advocate of MPAs...This raises the question of impartiality and the need to balance the advice from Roberts with advice from scientists who have not adopted a public advocacy role for MPAs. Fourth, the names of the "independent international scientists" who were commissioned by the Chief Scientists of Defra, NE and JNCC to peer review these three reports, have not been revealed, nor have their reviews been published' (Gray 2010).

JNCC (2010, p. 2) had made it clear that 'The priority in MCZ planning is satisfying the requirements of the Ecological Network Guidance (ENG)'. Keith Hiscock (2013), a marine biologist member of the SAP, criticised the ENG for being too rigid: 'When the ENG was "Guidance", it could be interpreted and the possibility of applying common sense was possible. But the guidance became a rule book and we are not dealing with the sort of scientific certainties that would allow such'. Hiscock (2013) complained that the ENG incorporated three 'flawed concepts for marine biodiversity'—connectivity; viable areas; and adequacy—and he referred to the 'foolishness of making rule-of-thumb guidelines for such poorly-understood concepts into rules. Desperate searching for areas of habitat…occupied by a species that satisfied ENG guidelines was also time wasted on trying to satisfy scientifically flawed concepts'. Similarly, ENGOs complained that the exercise was target driven rather than science driven:

> *there were difficulties in terms of people's understanding of what was being presented to them in terms of science…it was a case of picking out lots of different colours to meet targets…It never took an ecosystem approach; it never looked in detail at different data layers that present a picture of specific habitats…the focus was very much target driven. We need a bit of that habitat that happens to be that colour.* (ENGO manager)

Jim Evans, chairman of the Cardigan Bay Fishermen's Association, made similar criticisms of the Irish Sea Conservation Zones (ISCZ) regional project: 'At the ISCZ meeting I attended…When it came to site selection, we were just given coloured squares of card and invited to put them where we thought conservation sites should be situated. There was no science at all' (*Fishing News* 15.4.11: 6).

There were also concerns whether the ENG could be understood by some stakeholders. One static gear fisherman, when asked for his thoughts on the ENG, said: '*The ENG—Jesus Christ—makes me dizzy to sit down and digest it. It was their* [NE and JNCC] *brainchild, it was their criteria—all generated from their viewpoints, based on papers not even published yet. It was based on "in-house" science, that didn't just add up.*' Whilst this fisherman's anxieties over unpublished, in-house science may be exaggerated (the ENG cited numerous peer-reviewed sources from a special issue of the journal *Ecological Applications* (2003) that focused on the planning of a network of MPAs in California), his concern over the ENG being derived from an exclusive NP elite's view of science may be well-founded. A member of the NFFO challenged the way science was construed by NPs: '*I think I've got a problem with the word science. It's been corralled by biologists, whereas if you look at the work by Elinor Ostrom, the social and the economic sciences of resource management are every bit as valid'.*

Jim Portus (CEO of the South Western Fish Producer Organisation) summed up graphically fishers' fears over the objectives of the MCZ project: 'the fishing industry is being subjected to a huge experiment in the name of 'marine planning'. Such experiments in town planning gave us monstrous high rise blocks of flats…[these]

blots on our landscapes have since been razed to the ground. However, our poorly-understood seabed may be forced to suffer longer, pushing further into uncharted territory' (*Fishing News* 6.5.11: 8).

Controversy over Data Deficiencies

One of the most contentious issues in the controversy over MCZs was the lack of comprehensive data. Mosquera et al. (2000, p. 321) claimed that data deficiencies were a systemic problem for MRs: 'the models needed to assess the conservation benefits of marine reserves are often as complex as those used for conventional fishery analysis and cannot realistically be applied to many of the fish populations that are threatened by fishing'. MacMullen (2010) pointed out in the English case that 'According to the British Geological Survey, only 15% of our seabed has been properly mapped in respect of ground conditions, habitat types and species assemblages'. Lockwood (2013b), the retired government marine scientist and Chair of the MPAC, stated in a letter to *The Times* that 'Rarely, if ever, were quantitative data presented in preparing the English list of 127 proposals for MCZ…Mr Benyon is right to describe the unsubstantiated beliefs, assertions and hyperbole of the nature conservationist campaign as banal'. MPAC said that 'The whole question of the *quality* of the ecological science and information on patterns of fishing is an area of major concern for the Coalition to which to date, Natural England's assurances seem hollow. An exercise in gauging the sensitivity of different offshore activities to marine conservation zones has been abandoned as inadequate, after criticisms from participants…[which] does increase the sense of *making it up as go along in a hurry* and of a forced process' (*Fishnewseu.com* 31.8.10; italics in original). Lockwood took particular exception to unevidenced proposals for no-take zones: 'What we do not have from anywhere in the NE Atlantic region is evidence that non-specific, unfocussed NTZs as proposed among the 127 sites that Whittingshall is demanding—yield any benefit whatever…the Lundy NTZ has delivered no evidence of conservation benefits other than to lobsters. Yet if lobster enhancement had been the primary purpose, it could have been delivered by less draconian, more focussed fishery legislation' (Lockwood 2013c).

The NFFO caricatured the NPs' cavalier attitude to data: 'we don't need to understand the complex interrelationships within the marine environment and between it and people's livelihoods in order to act, and we must act in urgency before it is too late' (*Fishnewseu.com* 7.4.09). However, (as we saw earlier) despite considerable pressure, the government did not accede to these NP demands, but maintained its flexibility stance that 'the management measures needed for MPAs may vary widely depending on the objectives of each site and the sensitivity of the protected features to different activities and levels of disturbance. In some cases this may mean that only seasonal or time limited restrictions are required, if any. In others it could lead to complete restriction of activities on a site, which is commonly referred to as a highly protected marine reserve' (Defra 2006b, p. 101, para. 10.65).

MCZ data deficiencies manifested themselves in three ways: not knowing precisely the location of the conservation features that needed protecting; not knowing the conservation status of the features that needed protecting; and not knowing the impact of a given pressure on the conservation features that needed protecting. Given this triple uncertainty, documents preceding the draft Marine Bill stressed the need to adopt the precautionary principle to protect the marine environment (Defra 2004; Laffoley et al. 2004). This stance followed the position taken by the RCEP (2004). However, despite the emphasis on the precautionary principle in the documentation leading up to the Marine Bill, as Appleby and Jones (2012, p. 74, 76, 77) observed, the White Paper adopted an evidence-based approach, and the MCAA made no reference to the precautionary principle, which they claimed made a mockery of MCZ designations: 'The danger…lurks…that after a complex scientific process MCZs will be designated…only insofar as they affect no ongoing activity… paper designations would be a shameful result' (see also Jones and Carpenter 2009; Eades 2012).

The planning of English MCZs was based on the widely-known idea of 'systematic conservation planning'[15] (Smith et al. 2009), which required the protection of around 20–30% of each of the UK's broad-scale habitats[16] that represent its marine biodiversity, as set out in the ENG policy document (Ashworth et al. 2010). This approach relied heavily on modelled habitat data, and its adherents argued that even if the precise location of sensitive habitats was not known, it was likely that some vulnerable patches would be included in the network. Such an assumption, which implied that the ad hoc placement of MPAs would work just as well as scientifically-substantiated locations (Roberts 2000), was, however, hotly contested by other marine scientists (Dinmore et al. 2003; Kaiser 2004; Greenstreet et al. 2009; Abbott and Haynie 2012). The fishing industry was also very critical of the systematic conservation planning approach applied to MCZs: 'When in the majority of the planning areas there is very limited ecological data that identifies what is actually there to protect, you know that decisions cannot be robust and this calls into question what scientific purpose these areas could really serve' (Rodmell 2011). David Stevens, a Newlyn skipper, described NE's advice to government as 'more like science fiction than science' (*Fishing News* 29.1.10: 7). Another fisherman commented after a Net Gain meeting that 'If it's conservation they are worried about, it is the fishermen themselves who are the endangered species' (*Fishing News* 29.1.10: 7). A SC fisheries scientist respondent insisted that robust evidence of species at risk was needed before the closure of a fishery could be justified:

> *Say you were going to close off an area just because you think that it might be an area where there's a delicate species, or there are species that need conserving, but then that area is also an important area for a fishery. I think that you'd have a hard job selling it to the fishermen that they should close this area off if you couldn't actually point to real*

---

[15] This concept was originally used to plan networks of terrestrial protected areas (Margules and Pressey 2000).

[16] There are 26 broad-scale (EUNIS, European Nature Information System (level 3)) habitats in English waters.

reasons why you were going to do it, because they would say you're shutting us out of this area with no particular evidence of importance, so how do you justify that, and I don't think you could. If you're going to ask fishers to reduce their impact, then I think you've got to have good evidence that it's going to make a difference; otherwise you're essentially just arbitrarily closing them out from their living.

By contrast, Côté et al. (2001, p. 178) argued that the very lack of data favoured MRs over CFM: 'in areas in which a lack of scientific information precludes conventional population-based management, the establishment of MPAs simplifies management and reduces enforcement costs'.

The notion of protecting 'representative' habitats also raised concern: '*I think that there's a case for vulnerable habitats for measures. I think the case becomes much more difficult to argue when we're talking about representative habitats*' (Fishing industry representative). Without support from the industry, it is difficult to see how 'representative' sites could be identified, but the fishing industry stated that data on the distribution and intensity of fishing activity should not be used against them (NFFO 2009a), so whether it would be willing to provide data on fishing effort to assist the process of identifying representative habitats, is a moot point. Clearly, their data is regarded as a political resource that the fishing industry can use to bargain with. Network connectivity was also problematic because of data paucity. Richardson et al. (2006a, p. 27) stated that in their Defra-commissioned investigation of the evidence base for the Marine Bill, 'Network connectivity…was not addressed due to the complexities of the task, data inadequacy and the number of scenarios involved'.

The MPAC was particularly critical of the lack of data on the displacement effects of MCZs. Lockwood said that the government 'has hardly begun to scratch the surface of the displacement issue, yet it carries huge implications for the economic consequences of MPAs, as well as ecological degradation outside the protected zones' (*Fishing News* 24.12.10: 24). Rodmell complained that 'the displacement of…fishing activity from important customary grounds, without careful forethought can lead to a redistribution that increases pressures upon less resilient and more pristine habitats. The conservation gains achieved within an MPA can be more than offset by losses at an ecosystem wide scale…During the passage of regional projects, the panel [SAP] chose to ignore requests from the regional projects for its views on fisheries displacement. Similarly, the government's Statutory Nature Conservation Bodies which have been running the English MCZ project have refused to recognise displacement as a conservation issue' (*Fishing News* 13.1.12: 17). For the MPA Coalition, displacement was the major issue of contention in MCZ siting (Lockwood 2013a).

Another contentious issue over lack of data was the alleged damage caused by dredging and trawling, which NPs claimed justified prohibiting them in all MCZs:

the most damaging gears for my money are dredging and trawling and they've done immense harm both to the sustainability of the stocks that they catch but also the habitats that those stocks occupy, so I think we need to shrink the footprint of these mobile fishing gears by a lot…there are conservation benefits to be had from static gear-only areas…for me an MCZ that doesn't protect against mobile gears is not worth having, it will just be a paper park. (Marine scientist; see also Monbiot 2012)

These assertions stem from high-profile studies that have been critical of trawling as a fishing method (Watling and Norse 1998). However, whilst trawling has undoubtedly changed many UK seabed habitats (Jennings and Kaiser 1998), the fishing industry contested the severity of its impact. One ex-fisherman respondent said:

> *I suppose trawling causes damage...But...fishermen return to the same fishing grounds year after year and they catch fish there. So you have to ask yourself, how much damage does it really do? You could say it's almost like ploughing the land...The trouble with some of these environmentalists is that they've got this vision of returning the seas to the pristine state they were in before industrialisation. That's no more realistic a prospect than it would be to return the land to its pre-industrial state. We are where we are.*

Different habitat types have different sensitivities to trawling/dredging, and the fishing industry argued that a blanket ban on trawling in MCZs would be indiscriminate and disproportionate (NFFO 2012c).

One of the government's attempts to rectify the lack of data—the establishment of 'reference areas'—further revealed the deep gulf between the NP and SC perspectives. The government announced the introduction of no-take 'reference areas' (where all extractive activities are banned) as control areas to monitor MCZs. The Wildlife and Countryside Link justified them as necessary for scientific research, enabling comparisons to be made between sites subjected to little or no human impact (the controls or reference areas) with sites subjected to extensive human impact (Link 2011). But the fishing industry rejected reference areas as unnecessary: In many respects they represent the epitome of the conservation land grab, anti-people philosophy that has never been far below the surface amongst parts of the conservation lobby and patently within the government's conservation agencies which are prescribing such areas (Rodmell 2011).

In the event, Defra did not include reference areas in the first tranche of MCZ designations in 2013. Sue Wells (2014, p. 7) from NE explained that the reference areas were 'controversial, and the regional projects had great difficulty getting stakeholder agreement on those that were recommended. Defra decided not to include them in the 2013 designations following advice from the SNCBs that these [65] recommendations did not meet the requirements laid down in the ENG and that a further review of both the sites and process used to identify them is needed'.

Controversy Over Time-Scales

Linked to the controversy over data deficiencies surrounding the planning of MCZs was a controversy over time-scales. Several international regimes issued explicit deadlines for the designation of MPA networks. For example, the WSSD (2002) called for the establishment of MPA networks by 2012, and the legally binding MSFD (2008) incorporated this target. The OSPAR convention held that an ecologically coherent network of MPAs should be established by 2012, and that by 2016 it should be well managed (OSPAR 2010). The rush to establish MCZs was also partly due to the pressure applied on the UK government by the advocacy coalition of ENGOs (see Chap. 2). As the NFFO observed:

> A classic and largely artificial moral panic about the supposed imminent demise of hundreds of thousands of marine species, and the widespread collapse of commercial fish stocks, floated the Marine Act through parliament. It also led the Government into a rushed and deeply-flawed process of establishing a network of MPAs through a *big bang* process. Instead of an incremental, steady, approach where one MPA would be trialled and necessary lessons learned before going onto the next MPA, armed with that experience, we are in the middle of a headlong rush on all fronts at once. (NFFO 2010; italics in original)

The MPAC said that the SNCAs were 'driving the MPA agenda forward at breakneck speed' (*Fishing News* 19.2.10: 3).

An MP who had a fishing constituency noted that:

> *My impression of the process was that it was a rush of virtue to the head from those keen on marine conservation, but the fishermen weren't taken enough into account and haven't been generally. There is a kind of wildly misinformed view that the fishing industry is damaging the stocks, and damaging the seas, and damaging the environment and therefore we need to protect the sea from the fishermen, crazy!*

It has also been argued that conservationists cut corners when, having been campaigning a long time for a MCAA, they had to come up with MCZ proposals at very short notice: '*We have been campaigning for a decade for the MCAA to come through...In terms of the identification of the MCZs that has been much more rapid...and resulted in a lot of key decisions being made very late on in the process.*' (ENGO manager). Dave Cuthbert, an industry representative on the Finding Sanctuary regional project, said that 'The MPA network has taken nearly two years to plan and discuss with stakeholders—but after the next Finding Sanctuary meeting, Reference Zones are expected to be agreed after no more than two days of discussion and no real chance for stakeholder engagement. This will take place without any approval from the fishing industry reps whatsoever' (*Fishing News* 25.3.11: 4).

Some authors were critical of timetables because they were arbitrary and could result in designating MCZs in inappropriate locations (Wiersma and Nudds 2012). The fishing industry added two further reasons for delaying MCZ designations: the 'best available data' cannot be robustly analysed in a short period of time; and 'trust' takes time to build between stakeholders—trust being a factor which social scientists argued is essential for successful planning (Glenn et al. 2012). There was in fact a delay in the MCZ site selection process in 2011, as Benyon explained: 'it is clear from the SAP's advice that there are a number of gaps and limitations in the scientific evidence base supporting the MCZ recommendations…the need to strengthen the evidence base for the MCZ recommendations means this is going to take longer than the ambitious target first put forward' (*Fishing News* 25.11.11: 2).

Controversy Over Consultation

Many commentators emphasised the importance of stakeholder participation in MR decisions, on grounds that without substantial stakeholder involvement throughout the planning process, MPAs were likely to fail (Grorud-Colvert et al. 2011; Olsen et al. 2013; Agardy et al. 2011) As both Rees et al. (2013) and Jones (2012) pointed out, the MCAA prescribed stakeholder participation in the MPA process. James Marsden (Director Marine, NE) held that NE fully endorsed the notion of stakeholder consultation: MCZs were a 'new type of MPA which gives fishermen a real opportunity to have a seat at the table, get involved and contribute to decision making. Experience from around the world shows that the best way to establish MPA networks is in partnership with sea users and everyone else with an interest in the sea' (*Fishing News* 29.1.10: 4). Defra defended its efforts at consultation on the MPA issue, saying, for example, that it received around 40,000 responses to its request for feedback on the MCZ proposals via its website (*BBC News* 21.11.13), although Wells (2014, p. 7) from NE explained that 97% of these responses 'were generated by the campaigns calling for the establishment of the full network', indicating the hard work carried out by the pro-MPA advocacy coalition.

The four regional projects that Defra established involved extensive stakeholder consultation during 2009–2011. Each regional project had a regional stakeholder group which worked on MCZ site-selection, supported by the regional project board's scientific expertise. The proposals received from the four regional projects (for 127 MCZs and 65 highly protected reference areas) were assessed by SAP for their consistency with the ENG, after which SAP published its own recommendations in November 2011, revealing some data deficiencies in many of the proposals. NE and JNCC were then required to obtain further evidence to fill these data gaps, after which the recommendations were presented to Defra along with socioeconomic impact assessments, for its designations after a process of public consultation which was undertaken in 2013. Pieraccini (2014, pers. comm.) stated that 'participatory techniques have become key elements of the designation process of MCZs in England', while Fletcher et al. (2014, p. 265) claimed that 'The commitment to stakeholder involvement in the development of the MCZ network was unprecedented'. Even Lockwood (MPAC Chair) (2013b) said that 'the UK fishing industry has been fully and positively engaged with all UK administrations in the MCZ site-selection process'. Moreover, the government proposed to involve stakeholders in the implementation of the MCZs, by encouraging a partnership approach through the creation of voluntary agreements between regulators and marine users (*Fishing News* 29.11.13: 2).

Nevertheless, several voices complained that the stakeholder consultation process was flawed. One complaint was about the one-sidedness of the consultation on the Marine Bill: there was an overwhelming response to that consultation process from the green lobby (MCS, Wildlife Trusts, and RSPB) which contributed 74% (916) of the replies that the government received. In stark contrast, the fishing industry contributed only 1% of responses. Thirty percent of respondents made a case for MRs[17], with no user sectors opposing (Defra 2006a). This confirmed the

[17] As the MCAA was being drafted, a campaign organised by MCS, Co-operative Society and Public Aquaria (British Marine Life Study Society) collected more than 500,000 signatures from the public in support of MRs (http://www.mcsuk.org/mpa/mcs-position-statement).

fears of some members of the fishing industry that it had been too slow to engage in the consultation process on the content of the Marine Bill to make sure its interests were taken into account during consideration of nature conservation proposals. Another complaint was about the effectiveness of the stakeholder involvement in the proposed MCZ designations, as we saw earlier (Fletcher et al. 2014). Jones (2012, p. 248) noted that 'Whilst the initial processes by which MCZ recommendations have been developed provided for stakeholder participation (bottom-up), the main steer has been from central government (top-down). The subsequent designation and implementation of MCZs is likely to be more top-down'.

Pieraccini (2014, pers. comm.) described the designation consultation process as 'thin'. She said there was a risk of 'overstating the participatory character of the designation of MCZs…the whole process of designation occurred on terms initially set by the government within a pre-given ecological representation offered by the ENG and a set ecological goal' (Pieraccini 2013, p. 119). Hiscock (2014, p. 47, 48), who was a member of the SAP, said that the ENG was non-negotiable: 'In the MCZ process, the ENG was a "given"—it was to be followed religiously (indeed it was described as our "bible"). Evidence that we have now suggests that many of the "givens" in the ENG were flawed…In many cases, flawed "evidence" of what was where was accepted, perhaps to fulfil quotas for habitats and species'. Dale Rodmell (assistant chief executive of NFFO) who participated in three of the four regional projects, said that 'The approach through the projects continues to be heralded as stakeholder led but in fact sites were selected under duress imposed by a top-down policy document, the Ecological Network Guidance (ENG), that in many cases offered few, if any, alternatives if the guidance was to be fulfilled. Far from having a strong stakeholder mandate, therefore, many of the sites are deeply contentious. Yet the guidance itself amounts only to an enumerated set of theoretical principles' (*Fishing News* 13.1.12: 17).

NFFO also complained that the so-called 'independent' SAP was far from independent, but reflected a bias in favour of MRs. Noting that the panel of nine was composed wholly of marine ecologists and natural scientists, notwithstanding previous indications that it would also include experts in social and economic science, the NFFO said it 'raises fundamental questions over…how far natural science on its own should be allowed a free rein in designating MPA sites…the very public pronouncements of particular scientific advisors and their close links to the green lobby and their *cause célèbre* of MPAs as a saviour of the marine environment, calls into question their objectivity in dealing fairly with matters affecting the livelihoods of those dependent on the seas…[The SAP] looks much more like a thin scientific veil for what are in fact political decisions' (*Fishing News* 18.12.09: 5). In reply to this charge, the chair of the SAP (Dr. Peter Ryder) said that socioeconomic 'expertise exists within the regional projects where the network design task will be carried out' (*Fishing News* 19.3.10: 18). This statement was hardly likely to reassure the NFFO because there was no guarantee that socioeconomic expertise was available in the four projects. A further criticism of the SAP was that there was only one member whose primary expertise was in fisheries science—Professor Michel Kaiser. Rodmell commented that 'It seems a relatively small clique of eco-scientists and MPA advocates, having realised their MPA *cause de célèbre*, have been given the freedom to construct an elaborate policy vision virtually as a scientists' writ' (*Fishing News* 13.1.12: 17).

The MPAC also complained that the statutory environmental advisor to the government—NE—had overstepped its duties and taken on the roles of MPA advocate and fisheries manager: 'sometimes advisor to government, sometimes sounding and behaving like an environmental NGO, and often assuming to itself the mantle of management decisions' (*Fishnewseu.com* 31.8.10). The MPAC claimed that its complaint resulted in NE's withdrawal from these illicit roles: NE 'announced that it will play no role in the design, implementation, or enforcement of the management measures that will apply within marine conservation zones...once they have been designated. That will be left to the relevant authorities: Defra, Marine Management Organisation (MMO) and Inshore Fisheries and Conservation Authorities (IFCAs)' (*Fishnewseu.com* 31.8.10). This represented a considerable coup for the MPAC.

Controversy Over Equity

The fifth controversy between the NP and SC perspectives was over the equity or fairness of the MCZ site-selection proposals. Cinner et al (forthcoming: 2) explained that 'wealthier fishers are expected to benefit the most from marine reserves because they can often influence the marine reserve establishment process to better fit their needs, and are also better poised to take advantage of alternative opportunities...fishers entrenched in poverty may not have the resources or capacity to fish further afield or diversify livelihoods'. Christie (2011, p. 177) noted that 'all boats may...not rise equally with MPA implementation and rent capturing by influential social groups is likely without mechanisms for equitable distribution of benefits'. This is an important issue, since, as Agardy et al. (2011, p. 228) explained, 'Perceptions that an MPA or reserve unfairly singles out a particular user group can...affect compliance and potential for criminal activity...Lingering distrust can complicate or even derail MPA planning processes'. However, Spalding et al. (2013) pointed out that equity is explicitly included in the Aichi Target 11 as a requirement of MR site selection, and they claimed that it is increasingly being taken into account in practice. Defra confidently declared that 'It is not envisaged that any equity and fairness issues will arise as a result of Marine Bill policy proposals. In line with the principles of sustainable development, social, economic and environmental considerations will all be taken into account for any decision-making involved in the Bill proposals' (Defra 2008a, p. 23, para. 6.3). One reason for this optimism was the fact that MCZs were predicted to bring fisheries benefits. For example, a marine conservation scientist, when asked about the potential impact of MCZs on inshore fishermen, said that the long-term impact would be positive:

> *I think the inshore fleet will stand to gain a great deal from well-enforced and protected MCZs even if they don't currently believe it. The evidence from other parts of the world is that those artisanal vessels end up getting good local catches from good local protection. So I think their fears are unfounded.*

However, this focus on the expected fisheries benefits of MCZs ignored the severe and, some would argue, disproportionate, short-term and long-term negative impacts of MCZs on the activities of smaller fishing vessels[18]—impacts which were glossed over when the Marine Bill was being passed through parliament: *'Well they just fish somewhere else. We are not shutting down the whole coast by establishing these MCZs, it is just a matter of moving your fishing from one place to another, and there's no constraint on how much they can catch, they have the same quota to catch what they did before...there may be some effects on some people but the price is worth it'* (Marine conservation scientist). But the negative impacts on the inshore sector could be serious. For example, although a study on the impact of 12 months of closure at Lyme Bay[19] indicated that the impact on scallop fishermen's profits was marginal, it acknowledged that Lyme Bay fishermen had to work harder and use more fuel to maintain their profit margins (Mangi et al. 2011). The study also reported potential long-term negative impacts of the closure such as increased conflict between towed and static gear fishermen, and concern over whether the smaller area now targeted by dredgers would be able to sustain the current number of vessels on a full-time basis over the long term (Mangi et al. 2011; see also Rees et al. 2010).

On the other hand, Fletcher et al. (2014) claimed that, far from ignoring socioeconomic considerations, the government was careful to pay close attention to such considerations by vigorously applying the concept of ecosystem services. Jones (2012, p. 254) noted that the regional projects 'provided for socioeconomic priorities to be considered through the participation of stakeholder representatives, in keeping with Article 117(7) of the Marine Act. This has proved particularly important in order to avoid MCZs in areas of high socioeconomic interest, where alternative sites can be proposed as MCZs, provided the overall network still complies with the ENG'. Indeed, Pieraccini (2014, pers. comm.) stated that emphasis on socioeconomic considerations was a distinguishing feature of the MCZ process, differentiating the UK approach from other EU countries.

Procedural justice requires that decision-making processes accommodate all stakeholder perspectives. However, the fishing industry voiced concern that the balance was struck too much in favour of the NPs: *'I feel that there's a bit of an imbalance between the interests of the fishers and the conservationists...The conservationists are full-time people working on this all the time, they can often make their voice heard over and above fishers who after all are having to attend meetings...in their spare time when they've probably been at sea all day plugging away in their job'* (Fisheries scientist). This respondent widened the charge of inequity to question the very legitimacy of NPs as stakeholders:

[18] The English fishing fleet is dominated by under-10m vessels that are largely confined to inshore waters (<6 nm), with 5326 such vessels comprising 82% of the English fleet in terms of vessel numbers in 2011 (Elliott et al. 2012).

[19] Lyme Bay, situated in the English Channel in South West England, is considered to be one of England's most important areas for marine biodiversity (Hiscock and Breckels 2007).

> *I'm not against conservation, but I think that the conservationists don't have a direct personal stake in fishing, to them fishing is just the enemy. And to fishers of course, this is the inshore guys who are operating small boats, it's their livelihood, and if they are arbitrarily shut out of large areas where they fish, they could well have to give up fishing and change their whole lives. Whereas for conservationists, closing off areas is no skin off their nose personally, and they might even get promoted for having done a good job! And I think this sort of imbalance between the true stakeholders and those who claim to be stakeholders through their general belonging to the population of Great Britain is very unbalanced.* (Fisheries scientist)

For their part, NPs argued that the MCZ planning process was biased towards industry representatives:

> *I think the most at any one meeting would have been five conservation-orientated people with...anywhere up to 35 stakeholder representatives from an industrial/socioeconomic background...it is very clear that five people in the room saying one thing and 35...the other then the process is not fairly weighted...We've also had problems of NGO staff basically being threatened by other stakeholders in meetings, and the project team in fact telling NGO staff they don't want to hear from them anymore in meetings. So I'll generally say it's an unfair process with an awful lot of bias towards industry.* (ENGO manager)

Another complaint of inequity made by NPs was the disparity in funding between them and their SC opponents:

> *throughout this process you've got to bear in mind that although a lot of industry stakeholders were financially funded or offered expenses for their time or money incurred, NGOs have had to fund it another way. We...sought funding from charitable trusts to employ staff to do the role but it's been a very, very financially heavy process...we are a charitable membership organisation and that's not something that comes easy.* (ENGO manager)

6.3.3 The Outcome of the MCZ Site Selection Process

In the event, the result of the MCZ site selection procedures which was announced on December 2012, was that Defra agreed to consult on only 31 of the 127 sites recommended by Natural England and JNCC[20] (and none of the 65 highly protected reference sites), which represented a partial victory for the fishing industry because these 31 sites were less intrusive on fishing grounds than were the remaining 96 sites. The reaction of NPs was, unsurprisingly, caustic. They pointed out that 59 of the 127 recommended sites were judged by the consultation group to be at 'high risk', and among the 96 sites rejected by the government, 36 were rated in the high risk category. In April 2013, Professor Callum Roberts led 86 marine scientists to sign an open letter to the Prime Minister, David Cameron, condemning the government for not accepting all the 127 MCZs recommended by the SNCAs after an £ 8 million consultation process. Pieraccini (2013, p. 117) pointed out that 'It is questionable whether designating a mere 31 sites is going

[20] The 127 sites covered 37,164 km^2, representing 15.3 % of England's total marine area. Added to the SACs and SPAs, which represented 12.8 %, they would have brought MPA coverage up to 27.1 % of English waters, and 34.2 % of its inshore waters (Jones 2012).

to make a noticeable difference to the establishment of an ecologically coherent MPA network'.

But Richard Benyon, then Defra Minister, said that the scientific evidence for a larger proportion of the zones was 'just not up to scratch' (*BBC News* 21.11.13): 'Scientists tell me there is simply not enough evidence to support these proposals' (*Fishing News* 16.8.13): 'every effort has been made to ensure that the selection of sites for the first tranche provides environmental benefits but does not go beyond what the evidence will support and does not unduly compromise coastal development' (quoted in Pieraccini 2013, p. 117). Dr. Judith Clarke, chair of North Western Inshore Fisheries Conservation Authority (NWIFCA), in a letter to Benyon, endorsed his decision: 'NWIFCA agrees with your comment that the 127 campaign is banal because it appears to seek designation of sites regardless of evidence. The Regional Committees that selected the sites had access to very little scientific evidence' (Clarke 2013). However, Roberts dismissed the argument of lack of scientific evidence: 'this is just a misleading excuse. Other countries, such as the US and Australia have developed world-leading networks with no more scientific evidence than we have. The strong suspicion is that they have caved in to the demands of the fishing industry, thereby hijacking a proposal that was democratic and inclusive, turning it into business-as-usual capitulation to the strongest lobbyists' (quoted in McKie 2013; see also Hiscock 2014).

In November 2013, Defra reduced the number of accepted sites from 31 to 27: two of the four rejected sites were turned down either because they would have caused too much economic disruption or because their habitats did not justify designation. This decision provoked even more outrage among NPs such as Roberts, who ridiculed the government's claim that the economic cost to fisheries and ports of some of the other 100 sites would be excessive: 'It's bollocks. These MCZs will not put fishermen out of jobs: they will protect them in the long run' (quoted in Carrington 2013). However, the new Defra Minister, George Eustice said 'We very much see the new MCZs as the beginning and not an end', adding that consultations on two further tranches of MCZ would begin in 2015 (*Fishing News* 29.11.13: 2), though he sought to reassure the fishing industry that a MCZ designation does not 'necessarily mean it is automatically a 'no-take' zone' (*Fishing News* 13.12.13: 2); 'Done creatively and with some thought I think you can protect these features and reassure other users such as the fishing industry that you are not destroying their livelihood' (*Western Morning News* 21.11.13). Eustice also sought to reassure the NP lobby, pointing out that 'It is important to remember that MCZs are only one part of the jigsaw. Over 500 marine protected areas already exist around the UK' (*Fishing News* 29.11.13: 2). The 27 zones covered 9700 km^2, and together with the existing 30,000 km^2 of protected area, they meant that 9% of all UK waters would have some form of protection (though some critics dismissed the existing protected areas as mere 'paper parks' with little effective enforcement).

Eustice defended the government's phased approach to MCZs: 'I think it is the right thing to do this in phases…To start from 27 and build from that. If you were to dump the full 127…in one go there would be the danger they would not be able to put in place those management measures from day one' (*Western Morning*

News 21.11.13). This phased approach appeared to be a recent stance adopted by the government: in 2009 Defra seemed to envisage a largely one-off process: 'It is intended to designate MCZs, for each region, by late 2012, in order to meet our commitment under the Convention on Biological Diversity. It is likely that a small number of MCZs will be designated after this date, such as where the issues are particularly complex or controversial, where further survey work is required to confirm the site's value, or in order to fill gaps in the coverage of MCZs' (Defra 2008b, para. 8.13).

This result meant that the SCs had managed to ward off a more serious challenge from NPs in the first round of MCZ designations. The role of the MPAC had been crucial in this campaign, as the fishing industry acknowledged: 'The MPA Fishing Coalition has provided the fishing industry with a platform that has helped to ensure that idealism has been tempered by pragmatism and realism' (*Fishupdate.com* 22.11.13); 'MPAC is credited with having brought a sense of realism to the implementation of marine protected areas—and halted a headlong rush to establish protected areas on poorly considered grounds (*Fishing News* 28.2.14: 5). The NFFO (which provided the secretariat for the MPAC) expressed its satisfaction with the result: 'Today's decision confirms the Government has balanced scientific information on vulnerable habitats, with data on the socioeconomic consequences of applying management measures within each designated zone to find a solution which brings benefits to all' *(Fishupdate.com* 21.11.13). Likewise, Pickerell (2013), on behalf of Seafish, the UK seafood authority, said that 'The minister has chosen to adopt a careful, incremental approach in order to balance the efforts towards conservation whilst minimising the socioeconomic impacts'.

In 2014, Defra announced that it was currently working on the second tranche of 37 potential MCZ sites (*Fishing News* 9.5.14). A public consultation on its decisions would be launched in early 2015, followed by the designation of further sites in late 2015. A third tranche of MCZ sites was expected to be designated in 2016. However, in June 2014, Callum Roberts complained that none of the 27 MCZs designated in 2013 had yet received any new protection, and that the authorities were 'falling over themselves to reassure the fishing industry that the zones will be business as usual…this leaves us in a worse position than before the Marine Act was conjured into being. Before there was nothing and we knew it. Now there is the illusion of protection. The person in the street will think the sea is well looked after at last, but there is still nothing. This network is less than useless' (*The Guardian* 17.6.14). Indeed, Roberts claimed that 'there appears to be no intention to give them any meaningful protection. We are building the world's most comprehensive network of "paper parks". If the present course continues, it represents an expensive exercise in futility' (*The Guardian* 21.6.14). Joan Walley, chair of the House of Commons Environmental Audit Committee, also complained: 'The government must stop trying to water down its pledge to protect our seas and move much more quickly to establish further protection zones and ensure they can be enforced' (*BBC News* 21.6.14).

6.4 Conclusion

Although Jones (2012) argued that a balance was achieved between top-down and bottom-up approaches in UK decision making about MCZs, the early stage of the MCZ process was dominated by an epistemic community composed mainly of natural scientists (see Chap. 3) supported by a pro-MR advocacy coalition of green NGOs whose core beliefs were focused on ecosystem preservation and recovery. The evidence presented in this chapter suggests that the clear objectives of the planning process, the data needs, and the time, consultation processes and equity considerations required to achieve successful outcomes, were not properly considered before the planning of MCZs started, the process being driven more by NPs who publicised a narrative of collapsing fish stocks and the continued destruction of UK marine ecosystems. The outcome of the first tranche of 27 designated MCZs suggested that the SC perspective recovered much of the ground previously gained by the NP perspective, but the battle between the NP and SC perspectives is by no means over, and will be fought out over the fate of the remaining 100 proposed MCZ sites during the next few months and years. There are some signs that attempts are being made to reconcile the two opposed perspectives, but there are also signs that the division between them is hardening. For example, the NPs have opened up three new battlegrounds—SACs, SPAs, and both together (European Marine Sites (EMSs)). On SACs, a report in *The Observer* (9.2.14) indicated that, frustrated over the MCZ outcome, NPs have turned their attention to the UK's 107 marine Special Areas of Conservation (SACs) set up in 1992 under the EU Habitats Directive, and found that regulators were allowing scallop dredging and trawling to go on in them. The Marine Conservation Society (MCS) brought legal challenges which forced the government to ban bottom trawling in 2013 in some SACs. On SPAs, NE was reported to be investigating the possibility of designating as a SPA under the EU Wild Birds Directive a large area of the sea out to 7 miles from St Austell to Falmouth Bay in the southwest of England, in order to protect seabirds from being caught in fixed/drift nets (*Fishing News* 11.4.14: 7). On EMSs, Client Earth and the MCS have threatened to send evidence to the European Commission that the UK's EMS are not being managed in accordance with Article 6 of the EU Habitats Directive, thereby triggering EU legal proceedings against the UK government (*Fishing News* 11.4.14: 7; *Fishing News* 28.3.14: 2).

The next (concluding) chapter summarises the arguments presented in the book, and examines their wider implications.

Chapter 7
Conclusion

7.1 Introduction

This book is a critical analysis of the major controversy that has gripped marine governance during the last 20 years—the vexed issue of marine reserves (MRs). Much of the book has concentrated on assessing the credentials of one side of this controversy—the nature protectionist (NP) argument in favour of MRs—because it was responsible for the most extraordinary speed with which marine protected areas (MPAs) rose up the academic agenda to become the most discussed policy measure in the scientific fisheries literature across the globe in recent years. The book also analysed the growing opposition to the NP perspective in the shape of the social conservationist (SC) argument, which is not opposed to selective siting of MRs, but rejects indiscriminate or blanket designations of networks of MRs based on flimsy data. The SC perspective recognises the need for improved management of marine natural resources that may include the use of some MRs to restrict certain users to stop damage to a specific habitat or species or to allow a particular habitat or species to recover, but argues that there is a risk of a policy misfit if the fisheries and socio-economic contexts within which MRs are sited are not taken properly into consideration (Jentoft et al. 2007). In particular, SCs are concerned that the beneficial fisheries effects of MRs may have been exaggerated, over-generalised, or sometimes advocated to decision makers on claims unsubstantiated by robust empirical evidence, often juxtaposed with an over-pessimistic narrative suggesting the imminent collapse of global fisheries. SC advocates claim that a greater research effort has been devoted to MRs than to alternative management tools because of the influence exerted by NP advocates who have pushed MRs up the policy agenda and subsequently attracted funding from ENGOs and governments to carry out more research in order to increase the 'attractiveness' of MRs to decision makers. If so, SCs worry that MRs may have been designated in what they deem to be the wrong places.

In this final chapter there are two sections. The first section summarises the arguments deployed in the book. The second section examines the wider implications of these arguments.

A. Caveen et al., *The Controversy over Marine Protected Areas*,
SpringerBriefs in Environmental Science, DOI 10.1007/978-3-319-10957-2_7

7.2 Summary

In the first chapter, it was stated that there are two competing views on the efficacy of MRs for fisheries management: the NP perspective which sees an extensive range of MR networks as vital to protect biodiversity and fish stocks from overfishing, which has become endemic across the globe, and threatens the survival of many commercial species; and the SC perspective which sees some selective MRs as justified within a strictly enforced conventional fisheries management (CFM) regime which includes multi-use MPAs (MUMPAs), and restrictions on quota, gear, and landing sizes. The chapter also described the growth of MPAs globally.

Chapter 2 explained how and why the NP paradigm of MRs became so dominant in fisheries management during the 1990s and 2000s. One reason was that it benefitted from a widespread perception that CFM had failed to prevent decline in fish stocks and in marine biodiversity around the world, and that a radical new approach was needed. Additional reasons were the work of marine ecologists who formed an epistemic community dedicated to the idea of MRs, and the activities of the international environmental movement which established an advocacy coalition to promote it. We noted, however, that this pro-MR advocacy coalition was belatedly challenged by an anti-MR advocacy coalition, which slowed its progress.

In Chap. 3, a bibliometric analysis of the peer-reviewed literature on MR was carried out to find evidence of a pro-MR epistemic community and/or advocacy coalition. A social network analysis identified key scientists, and a citations analysis identified key papers. The analysis plotted the trajectory of the rise of academic interest in the topic between 1970 and 2010 and identified those scientists who had the most numerous MR publications, who were the most connected with others in collaborative research, and who were involved in MR advocacy. The analysis also calculated which papers had been most cited and in which journals they had been most published. Among the findings were that MR papers dominated the marine management literature—research into MRs far outweighed research into all the other topics of marine management—and that 90 % of scientists who led MR research were marine ecologists. This suggested that a disproportionate amount of marine research funding was spent on MR work, and confirmed that a pro-MR epistemic community (EpC) and a pro-MR advocacy coalition (AC) existed.

Chapter 4 dealt with the issue of whether the dominance of the MR paradigm had led to a bias in the peer-reviewed literature towards pro-MR studies, and if so, whether that meant that the science/policy advocacy divide had been crossed. A questionnaire was e-mailed to 200 expert MR scientists to investigate whether the pro-MR EpC/AC was reinforced by academic journal publication practices that favoured pro-MR papers, thereby blurring the borderline between science and policy advocacy. The results showed only a little evidence of bias by journal editors towards pro-MR papers, but while no scientists admitted to self-censoring because of a pro-MR bias, one respondent was fairly certain that such self-censorship had occurred amongst some of his peers. On the issue of whether pro-MR scientists had crossed the line between science and policy advocacy, and if so, whether it mattered, we found that while some scientists held fast to the linear model of the EpC whereby scientific findings were conveyed impartially to policy makers, others adopted the deliberative model of engaging closely with policy makers in a joint process of decision-making based on data that was incomplete, and on normative assumptions that were contestable. In assessing whether this was acceptable, we noted that where there was a consensus of scientific opinion, scientific advocacy provoked less controversy, but that to guard against group-think, there should be a requirement that advocacy scientists provide a range of alternative policy recommendations.

Chapter 5 examined the evidence in the peer-reviewed literature (310 papers) from 1970 to 2010 for the benefits of MRs. Proof that MRs benefitted marine biodiversity was compelling, but evidence that MRs benefitted fisheries was thin, while very little research had studied the socio-economic benefits of MRs. Most of the empirical studies focused on MRs rather than MUMPAs; on tropical/warm coral reef systems rather than temperate/cold soft habitats; and on snapshots rather than longitudinal analysis, with few control sites or before-and-after analyses. This literature review was critical of targets like International Union for Conservation of Nature (IUCN)'s aim for 20 % of global seas to be designated as MRs by 2020, which had flimsy scientific support; led to hasty siting of MRs in inappropriate locations; disproportionately affected fishers' viability; diverted resources away from more urgent measures; and encouraged a false impression that the problem of overfishing would thereby be solved. The review also showed that the literature was skewed in various respects; that the scientific data available was deficient; that MRs were species-specific, habitat-specific and therefore site-specific; and that the problems of enforcement of regulations were no less real in MRs than in CFM regimes.

In Chap. 6, a case study of the UK's Marine and Coastal Access Act (MCAA)'s consultation process on the designation of Marine Conservation Zones (MCZs) was investigated. In an analysis of documentary literature and 21 key-informant interviews, the contrasting NP and SC viewpoints were employed as lenses to interpret opposing perceptions on five issues raised by the planning of the MCZ network: objectives, data deficiencies, time-scales, stakeholder consultation, and equity. The outcome of the first tranche of decision-making saw only 27 of the 127 recommended MCZs being designated by the government, which suggested that the SC perspective had prevailed against the NP perspective, though two further rounds of decisions are expected during the next 2 years, the results of which are likely to increase the number of MCZ designations depending on the robustness of the scientific evidence available for them. If so, this would be a sign that attempts are being made to reconcile the two opposed NP and SC perspectives, though there are also signs that the division between them is hardening.

7.3 Wider Implications

The wider implications of the above analysis of the MR debate are fivefold: the role of politics, the role of advocacy, the role of stakeholders, the role of caution, and the role of reconciliation.

7.3.1 Role of Politics in the MR Debate

There are two opposing views on the role of politics in the MR debate, one negative and the other positive. The negative view regards political motivations as irrelevant, counter-productive, and opportunistic influences on MR decisions, which should be based solely on ecological considerations. Charges of political opportunism can be found pitched against both NP and SC camps. For example, Dulvy (2013, p. 359) saw it in the NP camp: 'MPAs are often the conservation of a political opportunity rather than any unique biological feature and rarely has sufficient science come into the planning…MPAs are alluring because there is no apparent need for science to guide their designation because the concept of ring-fencing or banking biodiversity is intuitive to anyone, hence easy to sell as the least-complicated "magic bullet" solution'. According to Austin Mitchell, fishers viewed NPs as politically motivated: 'Many skippers believe that fisheries management is driven by politics rather than by conservation or sustainability requirements' (*Fishing News* 6.11.09: 6).

Dunne et al. (2014, p. 1) saw political motivation in the NP camp in the way that the Chagos MR was designated in 2010: 'The declaration of a 640,000 km^2 "no-take" marine protected area (MPA) in the Chagos Archipelago in 2010 was politically driven, and was preceded by a failure to examine fully the scientific rationale for protection. The entire area was already a highly regulated fisheries and environment zone which had been subject to a well-regulated licensing system with no evidence of over-exploitation. The cessation of commercial fishing is unlikely to provide effective protection for highly migratory pelagic species, and reef fish stocks were already in excellent condition'. Marine Resources Assessment Group (MRAG), who concluded that establishing an MR in the Chagos Archipelago would confer no benefit to tuna, proposed a partial approach, whereby the MR would be restricted to an area out to 12 nm from the islands, with a total coverage of 52,270 km^2, but this compromise was rejected by David Miliband, British Foreign Secretary, as Dunne et al. (2014, p. 31, 32) explained: 'Its creation was the result of aggressive advocacy from NGOs and scientists, and a government anxious to claim a "green legacy"…the decision to announce the MPA was a hasty political whim of a UK Foreign Secretary who was about to leave office, against the advice of his own officials and of the British Indian Ocean Territory (BIOT)'s expert fisheries advisers…the Chagos MPA…defies the principles which require a proper examination of the available evidence, and instead is an example of one where intense conservation advocacy and political motivation have triumphed'.

Other commentators, however, have defended the Chagos decision on ecological and fisheries conservation grounds. For example, Jones (2011) claimed that scientists said the MR would greatly reduce the pressure on tuna, and save thousands of shark and stingray from bycatch. Similar remarks were made by Professor Charles Sheppard, former tropical and marine environmental advisor for the Foreign and Commonwealth Office's Commissioner for UK Overseas Territories, who also said that the Chagos MR was justified because 'we have the opportunity now caused by government interest in doing something, which may not recur if we put this opportunity off' (e-mail 26.1.10 to Coral List Server)—a statement that confirmed the importance of political opportunism. Koldewey et al. (2010, p. 1907) also revealed political opportunism when they stated that 'The current extent, distribution, size and spacing of MPAs globally are vastly inadequate, particularly for no-take zones…There are only a limited number of sites around the world where establishing a large no-take MPA is practical…and the Chagos/BIOT MPA…doubles the coverage of the world's oceans that are currently strictly protected'. Likewise, Mangi et al. (2010, p. 6) expressed the politically opportunistic remark that 'To find such a large area all within one jurisdiction is remarkably fortunate from a conservation perspective'.

Hiscock (2014, p. 48) saw political motivation in the SC camp in the English case when he reported the 'suspicion amongst many…that meaningful sites and attributes that should have been listed for protection were not included because of political and industry interference and was nothing to do with lack of evidence'. Kaiser (2005, p. 1196) saw political motivation in SC when he referred to pressure exerted on CFM managers from fishers: 'The political need to appease a desperate fishing industry has tended to push management decisions towards, and even beyond, the upper confidence limits for future allowable catches'. But Kaiser (2005, p. 1198) also saw (future) political motivation in the NP camp when he predicted its likely presence in pro-MR decisions: 'the scale of MPAs required to ensure sustainable fisheries of wide-ranging, long-lived species such as cod and plaice, may be both impractical and equally prone to the same political horse-trading that has neutered many current management systems'.

The positive view of politics in the MR debate rejects the idea that politics undermines ecology, arguing instead that politics is an inescapable and healthy part of the ecological controversy in the MR debate as in every other policy debate. For example, Sumaila et al. (2000, p. 756) stated that 'Establishing MPAs is like any other public policy decision. It is a political process where scientific knowledge may inform the debate and influence the outcome, but the decisions are taken elsewhere'. This school of thought asserts that both NP and SC camps are political ideologies first and scientific theories second; although they both invoke scientific arguments about ecology, those arguments are essentially surrogates in a political struggle for power between them. The NP side accuses the SC side of politicisation of science in the MR debate by using science to support the socio-economic interest of marine users (especially the fishing industry), while the SC side accuses the NP side of politicisation of science in the MR debate by using science to support the move to meet abstract targets set by the environmental industry (especially the green NGOs). But the fact is that in each advocacy coalition, science plays a supporting role, whereas politics plays the leading role. This should not surprise us; since, as Aristotle said, politics is the master science. The language of politics is the language of priorities, and Lubchenco et al. (2003, p. S6) recognised the supremacy of politics in acknowledging that 'Because there will always be opportunity costs to conservation, there is a limit to how much we can conserve'.

Several other writers have acknowledged that political rather than ecological factors dictate many MR decisions. For example, Halpern (2003, p. S117) said that 'decisions on the design and location of most existing reserves have largely been the result of political or social processes'. Likewise, Jameson et al. (2002, p. 1180) claimed that 'most MPAs are designed and located based on socioeconomic and political issues…and rarely account for the ecology of organisms to be protected' (see also Cộté et al. 2001). Similarly, Hattam et al. (2014, p. 269) said that 'MPAs are the result of social processes or transactions, involving dynamic interactions between individuals and groups. They also result from political and societal views on desirable states of the marine environment'. On the positive view of politics, such political decisions do not undermine ecology, but simply exemplify particular interpretations of ecology. For example, if socio-economic factors prevail, the SC interpretation of ecology is likely to have triumphed. This is not the devaluation of ecology for political considerations, but the healthy outcome of a political conflict between different views on ecology. So the role of politics is not a matter for regret, but a fact of civic life, a recognition of which helps us to understand more clearly the reality of the MR debate. For example, it provides an alternative reading of the Chagos case by showing that that case was not an instance of politics undermining ecology, but an occasion where in the political battle that took place between NP and SC perspectives, the NP perspective prevailed. The positive view of politics thus affirms that politics prevails in every case, but that the supremacy of politics in MR decision making does not necessarily mean that such decisions are wrong: Whether the outcome is 'right' or 'wrong' depends on one's point of view. The antagonism between the NP perspective and the SC perspective leads to a political battle in which each side tries to muster sufficient political support to defeat the other's political supporters. The outcome is necessarily a political outcome, but whether it is a good or bad political outcome in any particular case depends on whether the observer is sympathetic to NP or to SC.

Whether we adhere to the negative or the positive view of politics, we must acknowledge that the conflict between the NP and SC viewpoints is ultimately a political struggle. We can see this illustrated in the central issue of enforcement of MR regulations. As Mosquera et al. (2000, p. 321) pointed out, on this issue, politics appears to drive MR designations independently of ecological considerations, in that MRs are allegedly easier to manage than are CFM measures: 'one of the advantages of reserves is that they simplify management and reduce the enforcement costs for fish populations where little biological information is available…This is one reason why they are often favoured for conservation in developing countries.' Similarly, Metcalfe (2013, p. 115) said that 'no-take MPA networks are frequently established for political reasons, such as less complicated regulations and easier enforcement'. So the critical factor for the success of MRs in protecting biodiversity and fish stocks is not fine-tuned scientific analysis of the ecological complexity of the closed area, but preventing marine users from violating the regulations—i.e. a political problem. Moreover, exactly the same problem has been identified for the much criticised system of CFM: the reason why CFM measures failed was not because the measures were inappropriate, but because the politicians failed to enforce them (Kaiser 2005). On this view, if we can solve the enforcement problem, we don't need to shift from CFM to establish MRs; and if we cannot solve the enforcement problem, there is no point in shifting from a failed CFM system to a failed MR system. The ecology of MRs is the wrong focus of attention: We should be concentrating on the real (political) problem of enforcement rather than the virtual (ecological) problem of pristine-ness. And we are more likely to solve the problem of enforcement if we establish MUMPAs than (large-scale) MRs. But of course, enforcement is a means not an end: behind the political problem of enforcement lies the substantive ecological issue of which measure (MR or MUMPA) is more likely to be effective in protecting fish stocks. So the political struggle is not for power in itself, but for the success of a particular strategy for dealing with an ecological problem.

In this political struggle, two trends can be identified: first, NP prevailed politically until the early 2000s, but SC has fought a rearguard action and has begun to prevail politically since then (Pajaro et al. 2010). Marinesque et al. (2012) claimed that this shift began as early as the 1980s. Second, NP still seems to prevail politically in large ecosystems which are remote from human habitation especially in developing countries (including colonial areas where political mobilisation of user interests is more difficult because of the disaggregation of the international fishing industry compared to the worldwide reach of the environmental industry); whereas SC seems to prevail politically in smaller ecosystems near to dense populations especially (but not exclusively) in developed countries (where mobilisation of local user interests is easier, and SNCAs/ENGOs have less purchase).

On large MPAs, as Devillers et al. (2014, p. 18) noted, 'Globally, the emerging trend is toward very large MPAs in remote parts of the ocean with limited potential for extractive uses and distant from the most serious threats to marine biodiversity' (see also Spalding et al. 2013; Marinesque et al. 2012). For De Santo (2013, p. 144), from the SC perspective, this trend towards large-scale MRs was regrettable: 'Short-term gains from closing off huge areas to meet international protected area targets do not equate with long-term enforceability or environmental sustainability, particularly for populations dependent on subsistence fishing for their livelihoods'. For Devillers et al. (2014, p. 1), from the opposite NP perspective, this trend to create 'residual' MRs was also viewed with dismay: 'As systems of…MPAs expand globally, there is a risk that new MPAs will be biased toward places that are remote or unpromising for extractive activities, and hence follow the trend of terrestrial protected areas in being "residual" to commercial uses. Such locations typically provide little protection to the species and ecosystems that are most exposed to threatening processes. There are strong political motivations to establish residual reserves that minimise costs and conflicts with users of natural resources. These motivations will likely remain in place as long as success continues to be measured in terms if area (km^2) protected'. Devillers et al. (2014) pointed out that out of the 10,000 MPAs now in existence, ten of them account for 53 % of the total global MPA area, and if another five LSMPAs which are being proposed are added, that figure rises to 74 %. Devillers et al. (2014) recognised that in the long run very large MRs would benefit biodiversity, but not in the short run, yet the short run problem was acute. Devillers et al. (2014, p. 18) held that these large-scale remote MRs were political decisions not ecological decisions: 'Large and remote MPAs are in many cases the only way countries can meet, at minimal cost and political risk, their international conservation commitments'.

Table 7.1 Potential roles scientists can take in policy debates. (Adapted from Steel et al. 2004)

Role	Description
Report	Scientists limited to reporting results and letting others make resource decisions, the 'traditional paradigm'
Interpret	Scientists interpret scientific results so that others can use them. This is often expressed as a scientist's contractual obligation to funding organisations that the results will be 'translated' for non-scientific users
Integrate	Scientists work closely with managers to integrate scientific results into resource policies and decisions
Advocate	Scientists recommend specific policies they prefer or believe flow from their scientific findings
Make final decision	In the face of highly technical and complicated issues scientists make resource decisions themselves

7.3.2 Role of Scientific Advocacy in the MR Debate

The above discussion on the role of politics in the MR debate helps us to understand the role of advocacy in the MR debate. Controversy over policy advocacy by scientists has been ongoing in ecology and conservation for almost a century (Brussard and Tull 2007; Nelson and Vucetich 2009; *Conservation Biology* 2006 volume 20(3)). Scientists' involvement in policy making can range from simply reporting research findings to policy makers being fully responsible for making a decision (Steel et al. 2004), as set out in Table 7.1. Few scholars would argue against the view that there needs to be *some* involvement of scientists in policy making, but determining what this role should be is the controversial issue (Nelson and Vucetich 2009). One possible role is narrative telling. A recent anonymous editorial on the communication of climate science to the public argued that climate change is as much about the kind of world we want to live in (values) as about the measurement of rising sea levels (facts), so 'scientists must learn to tell stories rather than report cold facts' (*Nature* 26.6.14: 444). Linked to narrative telling, is another possible role for science—advocacy—which entails scientists recommending to policy makers specific policies which they prefer, or believe to flow from their research findings. There are two views on the MR advocacy role: the negative view that scientific advocacy is bad and the positive view that scientific advocacy is good.

The negative view that scientific advocacy is bad rests on six arguments. First, scientific advocacy for MRs risks being 'unscientific, simplistic or selective' (Kaiser 2004, p. 637), undermining the principle of scepticism that lies at the heart of the scientific method (Robertson and Hull 2003), and damaging the reputation of scientists for objective research (Lackey 2007). Second, scientific advocacy for MRs can lead to the politicisation of MR science, whereby research into MRs is geared to highlight their benefits and play down their deficiencies (see Chap. 4). Third, as a result, scientific advocacy could devalue the peer-review process (Pielke 2007; Rice 2011). Fourth, scientific advocacy for MRs could diminish the credibility of MRs in the eyes of marine users. According to Gleason et al. (2013), a key reason why the California MR project succeeded was that scientists adopted a neutral role of advice-giving, not a value-laden role of advocacy. Fifth, scientific advocacy has been criticised for misleading the media and through them, the public. Scientists' communication of MR research to the media and general public may contribute significantly to how the issue is framed and whether it generates a critical mass of public support in favour of MRs, causing government to act (Polacheck 2006). Inaccurate reporting and sensationalism from the media (Ladle et al. 2005), can result in over-simplistic solutions to a complex problem becoming part of popular discourse, which can be counter-productive, as Agardy et al. (2003, p. 363) warn: 'When MPA advocates make sweeping statements about the benefits of MPAs, expectations are raised in user groups…Striving to meet these often unrealistically high expectations then puts unnecessary pressure on MPA managers, threatens the continued existence of these MPAs, and even endangers future MPA designations'. Sixth, scientific advocacy for MRs may divert research from more salient topics (such as the vulnerability of different marine species and habitats to different fishing methods) to less salient topics (such as connectivity criteria for MR networks).

The positive view of scientific advocacy for MRs asserts that the purpose of scientific research is to benefit society, and if the findings of that research are agreed by a substantial portion of the scientific community, scientists have a right—even a duty—to push for those findings to be implemented (Marris 2006; Scott et al. 2007; Polacheck 2006). Moreover, scientific advocacy for MRs gets the issue into the public domain, and encourages a wider debate, which enhances democratic decision-making. Such advocacy does not necessarily compromise objectivity, according to a marine scientist respondent:

> *The problem is that those who are against advocacy think that any scientist who speaks about something is suddenly abandoning their scientific objective and principles because they are speaking out. But actually no, you can promote your findings and say that look we have found in this particular area, this is what the science says more broadly about the outcomes of the implementation of protection, and therefore we can frame this as a solution to some of the problems that we know exist in the oceans and that is a perfectly legitimate use of science in my view…I don't see it as lacking objectivity…if I was to suddenly start twisting the science around and saying well…ignoring all the contrary evidence, or being very much cherry-picking about the examples that I was using, that then goes from scientific and objective to simply being an advocate, that is unhelpful.*

Indeed, for this marine scientist, the alternative to (legitimate) advocacy is not *objective* decision-making but *ignorant* decision-making:

> *I feel strongly that if experts don't speak out on these things, who is going to decide on them and who is going to be able to judge the validity of the arguments? Well it's going to be non-experts and if the expert voices don't get across the information or the evidence…in a powerful or effective way we will make decisions that are based on less good evidence and on opinions. Although ultimately that may happen still, if the good science is out there, then the chances of bad decision- making will be reduced.*

Moreover, those who defend scientific advocacy claim that it is a fact of life, over which scientists have little control. Pressure exerted by various social groups on scientists to fulfil advocacy roles put them in the uncomfortable position where advocacy is unavoidable (Steel et al. 2004; Gray and Campbell 2008). The important question, therefore, is not *whether* scientists should advocate, but *how* they should do so (Nelson and Vucetich 2009).

There are four ways of resolving this conflict between the negative and positive view of scientific advocacy of MRs. The first way is the **bathwater** solution of saying that criticism of advocacy should not undermine the value of MR science (we should not throw the baby out with the bathwater). As Laffoley (2012) puts it, 'you can say that advocacy has gone ahead of science, but that doesn't mean that there isn't some very good science at the heart of the matter that proves the basic concept of no-take MRs, that they benefit biodiversity in multiple ways and that what you get as a result of MRs is much better than not having one'. The second way is the **pluralist** solution of saying that provided there are advocates on both sides, the public interest is not at risk. However, if one side commanded much greater resources than the other, questions of inequity could arise. The third way to resolve the conflict is the **transparency** solution of saying that scientific advocacy for MRs is acceptable provided a distinction is publicly maintained between science and advocacy. There may be a fine line between scientists seeking to impartially provide advice to policy-makers, and becoming issue advocates (Pielke 2007; Scott et al. 2007), but, according to FAO (2006, para 9.8), it is a line that must be kept visible: 'the public deserves to be able to distinguish members of the discipline giving neutral advice of quality, from those advocating for their client, their sponsors or their own values. Advocates have a societal role to play but the danger is in the confusion of the scientific and advocacy roles'. Some scholars argue that if scientists make their underlying values explicit when they advocate policies, then their advocacy is more acceptable to their academic peers (Nelson and Vucetich 2009)—scientists are also citizens after all. However, the distinction between 'objective facts' and 'subjective values' is often not clear cut, because scientific information can be interpreted in different ways, depending on one's underlying worldview (Hilborn 2007b). Indeed, scientists may not even be aware of the subjective values that underpin their 'objective' advice, and therefore unable to recognise when they have crossed the line between giving neutral advice and giving value-laden advice (see Chap. 6).

The fourth way to resolve the conflict is the **safeguards** solution of saying that provided there are safeguards in place to prevent outrageously misleading and egregious claims being made by scientists, advocacy is legitimate. One such safeguard is to draw on experiential knowledge from practitioners in the field, such as fishers' ecological knowledge (FEK), to monitor factual claims made by pro-MR advocates. Another safeguard is to make use of social science. For example, social science research conducted alongside natural science on MRs would ensure that the socio-economic implications of designations of MRs in particular areas would be documented, dampening down a tendency by ecologist advocates of MRs to ignore such implications (Higgins et al. 2008; Rees et al. 2013; Christie 2011). Another safeguard against dubious or overblown scientific claims is to draw on conceptual analysis. For example, NPs invoke the concept of 'ecological coherence' to justify MR networks. But what does 'ecological coherence' mean? In 2012, OSPAR (2013, p. 32) admitted that 'no specific definition for the term "ecological coherence" has yet been formally agreed upon internationally and only a few theoretical concepts and practical approaches have been developed for an assessment of the ecological coherence of a network of MPAs'. Moreover, on this view, ecological coherence is not an independent criterion, but an umbrella term encompassing other criteria, which are themselves conceptually contested. For example, 'viability': MPA advocates often claim that MPAs must be large to be viable, but Hiscock (2014, p. 47) challenged this assertion: 'we know that many rare or threatened habitats and species occur in small areas (perhaps as small as a metre across) and persist there over decades and probably centuries. Whilst identifying tiny areas for conservation may be impractical from the point-of-view of management, the concept that MCZs need to be a prescribed minimum size to be "viable" was flawed'. Another controversial concept is 'network', which Hiscock (2014, p. 47) again deconstructed: 'The word "network" has become entrenched in the language of policy advisors and policy makers but is meaningless for all but a few mobile species…Attempts to identify connectivity distances between MCZs become mired in meaningless heuristics' (see also Sale et al. 2005).

A further safeguard is to draw on ethical theory to clarify the normative assumptions that lie behind some 'objective' scientific advice. A frequent criticism of science advocacy is that the scientific norms of 'disinterestedness' and 'organised scepticism' may have been violated during research on MRs (Merton 1973). The fact that much MR science has been funded by ENGOs and green trusts that are sympathetic to the MR cause raises serious questions about the objectivity of knowledge on MRs that is being produced. Indeed, normative issues lie at the heart of every natural resource management policy. As Mazur (1981, p. 41) asserted, 'Many technical controversies are primarily disputes over political goals and only secondarily concerned with the veracity of scientific issues which are related to these goals'. Weible (2007, p. 111) reported on the process of establishing MRs in California that 'the conflict over MPAs appears to be driven by normative preferences to establish MPAs'. Ethicists help to decide when a scientific advocate is reporting personal preferences rather than research findings.

7.3.3 Role of Stakeholders in the MR Debate

Given the above discussions on the negative and positive perceptions of the roles of politics and scientific advocacy in the MR debate, not surprisingly we find negative and positive perceptions of the role of stakeholders in the MR debate. On negative perceptions, stakeholder engagement creates potential problems for the SNCAs and the UK government who have to meet European targets, because the greater involvement of resource users in decision making could lead to the triumph of local parochial self-interests, and this could undermine strategic nature conservation objectives. Jones and Burgess (2005) persuasively argued that the European Court of Justice is unlikely to accept a government's defence that EU targets were not fulfilled because they were not consistent with decisions that emerged from local participation processes.

Positive perceptions of stakeholders' role in the MR debate include the claim that there should be a shift from the idea that 'the science is right' to the idea of 'making sense together'. The idea that the science is right with its associated nostrum that science speaks truth to power, depends on the highest quality of evidence. But in practice, such evidence is rarely available, both because of the lack of time and resources required to collect the massive amount of data necessary, and because of the difficulty of predicting the reaction of human and animal behaviour to management interventions (Fulton et al. 2011). An alternative approach to this linear model which privileges the place of the scientist, is the deliberative model of developing understanding together, which involves wider civil society in the production of knowledge. Non-scientists can focus on where key information gaps lie (e.g. contributing traditional ecological knowledge (TEK)), and can collaborate with scientists (e.g. in fisheries-science partnerships) which will improve the knowledge base and increase the legitimacy of scientific information for policy (Sweeting et al. 2011).[1] These recommendations draw on the concept of 'post-normal science', which acknowledges that in addition to systematic 'scientific' knowledge, there are other sources of potentially useful information for planning, including local knowledge and unpublished research that can fill critical gaps and allow decision-makers to make a more informed judgement than would be provided by a formal scientific assessment (Funtowicz and Ravetz 1991). In effect, this is a shift away from the deficit model of public understanding whereby scientists are 'educators' of the general public, to the engagement model whereby scientists debate, listen and learn with the public, collectively solving problems with them instead of imposing solutions on them (Jensen and Holliman 2009).[2] Such a shift towards more deliberative models of policy making acknowledges the messiness, complexity, and conflictual nature of marine resource problems (Stern 2005), and recognises that it is impossible for any single group or agency to possess the full range of knowledge needed to solve them (Berkes 2009). Essentially, post-normal science is a model of decision-making that is pluralistic and encourages open communication between resource users, conservationists, managers and scientists. Reflecting the idea of communicative rationality (Habermas 1984), open communication between stakeholders increases mutual understanding and trust, and improves the chances of successful policy outcomes (Hoefnagel et al. 2006).

Applying these nostrums to the MR issue, Mangi et al. (2010, p. 15) asserted that 'For over two decades, the idea that biodiversity can be conserved without considering stakeholders' interests, needs and aspirations, has been recognised as untenable'. Stakeholders not only contribute valuable knowledge to the MR debate, but may also suggest valuable MR initiatives. Creative solutions to a problem can emerge when scientists, policy-makers and resource users engage in a multiple-way dialogue with one another (Jasanoff 2006; Pielke 2007). An example is the MR, Cabo Pulmo National Park in the Gulf of California, Mexico, which after 15 years of protection, showed an average biomass that was five times greater than in fished areas in the Gulf of California—a success which Aburto-Oropeza et al. (2011) attributed in large part to stakeholder engagement. Kareiva (2006, p. R535) argued

[1] Though, the inclusion of wider society in the production of scientific knowledge may have implications for quality control (Funtowicz and Ravetz 1991).

[2] However, a Royal Society survey concluded that 'there was concern that many scientists still see the main reason for engaging with the public as the need to "educate" them rather than to debate, listen and learn as part of a genuine dialogue' (Royal Society 2006, p. 14).

that 'Far more important than modelling the ideal design of marine protected areas or networks of marine protected areas is building local social and community support for them'.

The traditional model of scientific expertise associated with central government has become discredited (Parry 2009) as stakeholders are wary of the credibility of advice that is offered to decision makers. For example, the approach taken in England to create an ecologically coherent network of MCZs was criticised as much for being too top-down and elitist as for its lack of substantive content (NFFO 2012b). For many people in the UK fishing industry there was a rejection of the process for establishing particular MCZs, rather than a rejection of MCZs per se. The persistent notion held by NPs that the planning of MRs should be natural science-led reflected an attitude that only scientists can speak truth to power, and ignored the fact that the democratic element of decision-making during the planning of MRs is vital to the MRs achieving their objectives (Voyer et al. 2012). Hilborn et al. (2004, p. 202) stated that 'Many countries have attempted to impose top-down catch or size regulations on local fishermen with little success. Top-down imposition of reserves is equally unlikely to work; what is needed…is bottom-up support of fishery stakeholders and communities' (see also Rees et al. 2013; Blount and Pitchon 2007). The top-down imposition of an MPA can also be costly for enforcement (Hanna 2003).

These considerations have led some observers to recommend the devolving of some MR decision-making from elites to stakeholders (Frid et al. 2006). Others have recommended that MRs should be co-managed (Jentoft et al. 2007; Jones et al. 2013). Co-management involves the sharing of power between central government and local resource users and managers. Austin Mitchell urged a partnership approach: 'The only management regime that will work is one which engages fishermen in the conservation process, rewarding them for good behaviour, rather than starting from the false premise that they are the root of the problem…A partnership approach to conservation has to be the way forward' (*Fishing News* 6.11.09: 6). In England, an example of local fisheries co-management is the system of IFCAs, some of which manage types of MR. Having local stakeholders actively participating in regional and local decision-making forums increases the legitimacy of a decision, while the central government acts as a co-ordinator, to ensure that national policy objectives are achieved across different regions, and as an adjudicator, ensuring that conflicts which are unresolvable at the regional/local levels are resolved at the national level (Phillipson 2002). Some of the more ardent advocates of the co-management approach argue that stakeholder participation should be viewed as a process through which objectives and actions are not settled in advance but emerge from the act of participation itself (Habermas 1984; Goodwin 1999).

7.3.4 Role of Caution in the MR Debate

Another important implication of the analysis of the MR debate in this book is over the role of caution. Both sides of the debate had significant things to say about caution: NPs urged policy makers to take account of the precautionary principle

(PP) in not allowing lack of data to prevent designations of MRs; SCs urged policy makers to exercise caution in the speed at which, and the extent to which, MRs are designated. On the NP stance, a striking characteristic of NP is that it readily invokes the PP where data are thin (Mosquera et al. 2000). NP scientists argue that the paucity of scientific data is not a sufficient reason for declining to designate MRs (Abdulla et al. 2009) because by the time the data become available, irretrievable damage may have been done to the relevant area. In the English case study, before the Marine Bill was drafted, Defra (2006b) emphasised the importance of the PP to protect the marine environment (Defra 2004; Laffoley et al. 2004), although the subsequent MCAA did not refer to the PP (Appleby and Jones 2012). So it appeared that NP lost ground to SC during the preparation stages of the Act. Another NP nostrum on precaution was their appeal for patience in monitoring the effects of MRs, because benefits may take a long time to materialise: 'Although some changes happen rapidly, it may be decades before the full effects of a marine reserve are evident' (Lubchenco et al. 2007, p. 6).

By contrast, SCs were very critical of the PP. For example, in the English case, several MCZs were proposed by scientists on the basis of limited evidence of the presence of a habitat/species vulnerable to fishing. The scientists justified their stance on grounds that imperfect knowledge was reason enough to invoke the precautionary principle so that restrictions were placed on potentially damaging fishing methods (e.g. dredging/trawling). However, the SCs (especially the fishing industry) claimed that the absence of robust evidence in support of placing a restriction on fishing was reason enough to allow current fishing to continue. Neither side could legitimately claim the backing of science, because the dispute was over competing (subjective) attitudes to risk, not competing claims of (objective) scientific evidence. SCs argued that invoking the PP was a tacit admission by NPs that the decisions were based not on science but on normative judgements, and that if MR decisions were normative, all stakeholders, not only scientists, should share in contributing to them. This confirmed the SC view that MR decision making cannot be only about the science (the linear model): it required an inclusive stakeholder process that encouraged people with different perspectives and interests to engage with one another to navigate trade-offs over the use of the marine environment (the deliberative model) (Salomon et al. 2011). SCs preferred to use the concept of the 'precautionary approach' instead of the PP: unlike the PP, the precautionary approach takes into account the cost effectiveness of implementing a management measure, so it requires that if the PP were to be invoked, it would not be applied in a way that inflicted disproportionate socio-economic cost on fishers.

But the SCs' own favoured tenet was the cautionary principle, by which it meant that MRs should be introduced with extreme caution. The phased and piecemeal outcome of the controversy over MCZs in England whereby 27 out of 127 recommended MCZs were accepted by the government in the first instance with a view to determining the configurations of marine use and reference area status within them after further investigations, is an exemplification of this SC viewpoint. Such a gradualist strategy enables marine managers to monitor the progress of early MCZs before embarking on further designations, giving more time for fishers and other marine users to adjust their practices to take account of the restrictions, and providing greater opportunities for consultation with stakeholders before the next rounds of proposed designations. Instead of adopting an arbitrary target of 10 or 20 or 30% of the seas to be designated as MRs, this phased approach is therefore one of adaptive co-management—i.e. putting experimenting and monitoring and consulting before acting: a practical example of learning by doing. It means abandoning the notion of a 'quick fix', with its twin myths of an immediate solution and a universal panacea.

On the first myth—an immediate solution—there are few if any situations in marine management where a solution can be found which is instantaneous. Invariably it takes time for measures to take effect, and even more time for their effect to be properly monitored and assessed. Babcock et al. (2010, p. 18260) pointed out that 'it will take decades to observe, predict, and validate the full implications of marine reserves'. The temptation to go for a quick fix is understandable, because scientific advocates, the media, and even the public on occasion, are impatient to see results. But that short-termist temptation should be resisted because it is unlikely to be as successful as a strategic planning and capacity-building approach that is incremental and process-orientated, rather than focused on the quick implementation of technical measures to meet pre-specified targets (though there may be statutory targets that have to be met eventually). The MPAC said that 'A big bang approach might be attractive from a PR perspective but increases the risk of getting it wrong' (*Fishing News* 25.11.11: 3). Moreover, there may be stakeholder resistance to wholesale designations of MPAs imposed too quickly without adequate consultation. So the success of MRs may well depend on the speed with which they are introduced.

Gleason et al. (2013, p. 97) explained that this was the approach taken in the successful California MPA project: 'the phased approach yielded positive outcomes for each successive region, as the Initiative team tested and learned which process design and outreach approaches were more likely to be effective'. Moreover, adaptation includes flexibility in re-siting MPAs which have lost their original rationale. For example, as Stelzenmuller and Pinnegar (2011, p. 181) noted, climate change may result in a species shifting its location to cooler waters, which 'can have consequences for the effectiveness of MPAs since the "centre of gravity" of a species' population range may shift gradually until it lies outside the boundaries of the original protected area' (see also McClanahan et al. 2008).

On the second myth—a universal panacea—Ballantine (2014, p. 4) seemed to assert that MRs *are* a universal panacea: 'The potential benefits of marine reserves are universal in scientific and social terms. They are independent of bio-geographical region and ecological habitat, and also of culture, politics and economics. Marine reserves can work anywhere'. Similarly, Roberts et al. (2005, pp. 129–130) said that 'The conservation values of marine reserves are universal—even blue-water pelagic habitats can benefit from protection from fishing…We have now learned enough about marine reserves to know that they have value for fisheries across the globe, regardless of geographical, political or management setting'. However, there are few if any situations in natural resource management where a one-size-fits-all measure exists that will solve all the problems: as Lubchenco et al. (2003, p. S3) pointed out: 'marine reserves are powerful management and conservation tools, but they are not a panacea' (see also Hilborn et al. 2004; Hughes et al. 2006). Kaiser (2005, p. 1194) said that 'they are not the cure-all that some purport' (see also Agardy et al. 2011).

In nearly all situations, MRs have to be customised and adjusted to meet the particular circumstances of different cases (Rudd et al. 2003). Defra explained that this was its approach to MCZs in England: 'to ensure effective, tailored protection for each of the sites' (*BBC News* 21.6.14). As Hilborn (2004, p. 197) explained, 'Their successful use requires a case-by-case understanding of the spatial structure of impacted fisheries, ecosystems and human communities'.

The point is that MR issues are complex, not simple: context is everything: 'Planning and implementation of ecologically connected networks of MPAs is context-dependent' (Kirlin et al. 2013, p. 11). The real estate agents' mantra of 'location, location, location' is also the key to MR siting: 'the impact of an MPA varies according to species' (Boncoeur et al. 2011, p. 192; see also Gerber et al. 2005; David 2005; Stelzenmuller and Pinnegar 2011). Austin Mitchell made the same point: 'our industry supports the creation of MPAs on a case-by-case basis: each requiring an evidence-based argument, clear objectives and a management plan to deliver them. We reject the notion that any MPA is a good MPA. We cannot accept arguments that merely seek the designation of some percentage of the seabed' (*Fishing News* 6.11.09: 6; see also Sweeting and Polunin 2005; Marinesque et al. 2012; Dulvy 2006). To adapt Edmund Burke, 'circumstances give to every MR its distinguishing colour and discriminating effect'. The appropriate mode of MPA management varies with circumstances, because every MR is unique, as Agardy et al. (2003, p. 356, 357) pointed out: 'The diverse array of MPA goals, and their order of priority, varies enormously from place to place—so much so that one could almost say that every MPA is unique, having been tailored to meet the specific circumstances of the place where it is established…Because specific circumstances vary so widely around the world, no model for MPA management objectives will be universally applicable… Such objective setting should be done with scientists working in concert with local communities, user groups, and management authorities—not by scientists in isolation' (see also Sweeting and Polunin 2005).

This phased, incremental, customised, flexible, piecemeal, case-by-case approach to MRs adopted by SCs was opposed by NPs who claimed that planning the location and size of individual MPAs on a site-by-site basis with the detail proposed by the fishing industry would seriously undermine their utility (Ardron 2008a; Smith et al. 2009), foregoing the widely-cited benefits of MPA networks (e.g. McCook et al. 2010). NPs argued that it was necessary to shift away from such 'piecemeal' approaches towards wholesale protection (Katsanevakis et al. 2011). On the other hand, Roberts et al. (2005, p. 125, 127), who are leading NPs, said that 'sites for protected areas should be chosen with care…Marine reserves need to be scaled appropriately for the species, habitats, and fisheries they are designed to support'.

7.3.5 Reconciliation Between NP and SC

Despite the normative gulf between NP and SC, five attempts have been made to reconcile them empirically—prioritisation, complementarity, compromise, integration, and morphing. The first attempt at reconciliation—**prioritisation**—provides that in situations where NP and SC clash, one takes priority over the other. For example, Rice et al. (2012, p. 218) pointed out 'the IUCN requirement that for any area to qualify as a protected area, "In case of conflict nature conservation objectives will be the priority"'. So if they clash, NP trumps SC. Conversely, many fishers hold that if biodiversity objectives conflicted with fishers' livelihoods, the latter should prevail. An intriguing example of prioritisation is implied by Sciberras et al. (2013, p. 3) who synthesised data on MPAs to compare the biological effects of three scenarios: (1) NTMPAs; (2) MUMPAs; and (3) open access. Their conclusion was that 'no-take reserves provide some benefit over less protected areas… [but] that partially protected areas are a valuable spatial management tool particularly in areas where exclusion of all extractive activities is not a socio-economically and politically viable option'. In other words, while NTMPAs were *theoretically* the most beneficial, MUMPAs were *practically* the most beneficial. (SCs might reflect that this signified that the best was the enemy of the good. They might also reflect on Edmund Burke's nostrum that if a policy was not good in practice, it could not be good in theory, since theory is necessarily about practice). Another example of prioritisation is implied by Stefansson and Rosenberg (2006, p. 67) who claimed that, economically, MPAs were better than failed CFM, but not better than successful CFM: 'the use of MPAs alone is economically inferior to a quota system which fulfils its promise of controls…but an MPA will be economically better than a quota system which consistently results in overfishing. It follows that there are conditions under which one system or the other may be preferred'.

In the second form of reconciliation—**complementarity**—the empirical gulf between NP and SC is seen to be not as large as is sometimes assumed. Rice et al. (2012, p. 217) reported that in a workshop of 100 participants equally balanced between SCs and NPs, it was concluded that 'although fisheries managers and biodiversity conservation agencies may give differing and sometimes opposing weights to the many objectives that could be set for MPAs, only 25 % of fisheries objectives and 30 % of biodiversity objectives were considered to be potential sources of conflict'. In other words, although the ethical priorities of NPs and SCs may be different, most of their practical objectives are common: 'Objectives for both fisheries management and biodiversity conservation have common goals of sustaining habitats and resources' (Rice et al. 2012, p. 228).

Another form of complementarity is the claim that NP and SC co-exist. Or rather, NP and SC *can* co-exist, if MPAs are designed in particular ways (there is nothing automatic about co-existence). Gladstone (2007, p. 72) implied co-existence when he stated that many of the MPA selection criteria for biodiversity and exploitable species benefits overlap, and 'the results of modelling studies, suggest that selection of MPAs for fisheries-related reasons may provide biodiversity benefits and vice versa'. Halpern and Warner (2003, p. 1871, 1877) also seemed to enunciate this form of reconciliation when they argued that 'recent findings in marine ecology suggest that this debate is largely unnecessary, and that a single general design of a network of reserves of moderate size and variable spacing can meet the needs and goals of most stakeholders interested in marine resources…it is entirely possible that most stakeholders could be served, and served well, by a common design for a system of marine reserves'. However, these statements are less a reconciliation of NP and SC than optimistic assertions that NP encompasses SC.

A more genuine form of complementarity is the argument that MRs cannot function in isolation but need CFM to regulate the spaces around them. For example, Sweeting and Polunin (2005, p. 3) stated that 'MPAs are not isolated from wider conditions. As spatially defined static entities, MPAs are vulnerable to environmental changes, including altered spatial and temporal distributions of fish and habitat, pollutants and eutrophication…For strongly-protected MPAs, other additional fisheries management measures are essential (e.g. large fleet or quota reductions)'. Similarly, Roberts et al. (2005, p. 123, 124, 128) stated that 'Fisheries management measures outside protected areas are necessary to complement the protection offered by marine reserves…Marine reserves are a powerful management tool, but work best if they are a supplement, not a substitute for other instruments…reserves complement, but do not conflict with, the great majority of existing management tools'. As Hale (2014, p. 5) put it, 'Islands of protection cannot exist within oceans of degradation' (see also Mora 2011; Agardy et al. 2011; Graham et al. 2011; Greenstreet et al. 2009; Halpern et al. 2010; Metcalfe 2013; Lubchenco et al. 2003, 2007; Hughes et al. 2006; Sumaila et al. 2000). In other words, NP needs SC. Similarly, Kaiser (2005, p. 1197) said that 'If the use of MPAs is to be successful, we first need to achieve effective fishing control. The failure or success of the use of an MPA as a fisheries management tool…is inextricably linked to effective fishing effort control in the surrounding waters'. Even Ballantine (2014, p. 4) accepted this complementary argument: 'Marine reserves are additional to detailed and general marine planning and management which will continue to operate outside the reserves'.

Conversely, SC needs NP. For example, SC needs some NTMPAs in order to protect fish spawning grounds and juvenile areas, protect biodiversity, and facilitate research and education. Roberts (nd: 2) stated that 'The question to be answered is not whether MPAs would be any easier to implement than present management tools, but whether they would be effective in supporting those tools'.

So NP and SC are mutually dependent, as Agardy et al. (2003, p. 359) argued: 'sustainable use on the one hand, and protectionist approaches embodied in no-take provisions on the other can complement each other for successful management'. Likewise, Jones (2014, p. 47) referred to a 'détente' between the two perspectives whereby 'both CFMAs and MPAs, including no-take MPAs, were recognised as contributing to the rebuilding of global fisheries and the related recovery of marine ecosystems…CFMAs and no-take MPAs are complementary…one cannot be effective without the other as both have their limitations'. Hale (2014, p. 5) asserted that 'We need all types of MPAs. There is not a single solution to protecting the ocean. We need multiple types of MPAs—from MPAs designed and managed to deliver specific ecosystems services to growing coastal populations, to very large, remote MPAs that act as baselines'. Perhaps NTMPAs could be used for biodiversity protection, while CFM and MUMPAs could be used for fisheries management? On this view, NP and SC can simultaneously co-exist (Jones 2007). The NFFO (nd) seemed to enunciate a complementarity model of reconciliation when it said that 'conservation and human utility of natural resources are not a zero sum game'. In other words, both ecological and socio-economic goals can be advanced simultaneously.

The third form of empirical reconciliation is a **compromise** between NP and SC—in particular, a compromise between the ecological requirements of NP and the socio-economic requirements of SC. Phil MacMullen (Head of Environmental Responsibility at Seafish) (2009) asserted that 'We must balance the need to conserve with the need to produce in a world that is becoming increasingly dependent upon marine resources'. For many researchers, striking the right balance between ecological and socioeconomic objectives is the aim (Lester and Halpern 2008). It is all about trade-offs: 'decisions to protect biodiversity inevitably come with trade-offs. Many of these trade-offs are social in nature…Before the MCZ network comes into force, honest discussion of the potential implications is therefore needed, in part to identify what trade-offs are being made, but also to identify which trade-offs are acceptable to the affected stakeholder groups' (Hattam et al. 2014, p. 277). One such trade-off described by Lubchenco et al. (2007, p. 15) was MPA networks rather than large-scale MPAs: 'A major socioeconomic benefit of a network is that fishing and other human activities can occur between the reserves instead of being excluded from one large area'. In the English MCZ case, the MPA Fishing Coalition conceived its role as achieving a compromise: 'our aim is to arrive at an accommodation that provides protection for vulnerable marine habitats…but in a way that allows fishermen to keep fishing. This is about finding the right balance between the nation's food security and protecting biodiversity' (*Fishing News* 26.2.10: 1). Defra also used the language of compromise when referring to the second tranche of candidate MCZ sites: 'Selection of sites will be consistent with principles used to select the first tranche, i.e. only those sites that achieve an appropriate balance between the likely social and economic costs and conservation benefits…proposed for designation' (*Fishing News* 4.4.14: 5).

The fourth form of empirically reconciling NP with SC is **integration**: fusing together some aspect of each theory. One example of integration is to recognise that there is considerable common ground between NP and SC in that both want healthy seas. NP is obviously committed to healthy seas, but so is SC, as Nikki Hale (chief executive of Eastern England Fish Producers' Organisation) pointed out: 'We should not forget that fishermen have got the biggest vested interest in protecting the marine environment' (*Fishing News* 19.2.10: 3). Similarly, Lockwood said that 'As with the conservation sector, the UK fishing industry wants a long-term sustainable future. It has no desire to be associated with short-term "rape and pillage" of the marine environment, as is too often suggested' (*Fishnewseu.com* 15.6.10). More explicitly, Roberts et al. (2005, p. 128, 129, 130) argued that 'Marine reserves… should be integrated into the fishery management toolkit because they can achieve things that other tools cannot…There is no surer way of integrating ecosystem level concerns into fishery management than protecting entire, intact ecosystems… Stocks of key fishery species in Europe, such as cod…and hake, need not have continued declining if juveniles had been protected from bycatch, since several excellent episodes of reproduction occurred while adult stocks fell. Integrating marine reserves into the management portfolio could have provided the necessary protection…Only when we add this tool [MRs] to fishery management strategies will conservation and fishery goals become completely allied'.

The fifth form of empirical reconciliation between NP and SC is that of one **morphing** into the other. SC may morph into NP if NP is seen as providing the long-term benefits sought by SC. Lubchenco et al. (2007, p. 12) enunciated this viewpoint: 'Although ecological goals often are viewed as being in conflict with some social and economic goals, recent research suggests that the choice is not between environmental and economic goals but rather between short-term gain and long-term prosperity. Long-term gains depend directly on healthy and resilient ecosystems'. According to this reconciliation, only a myopic or short-termist interpretation of SC is at odds with NP: a long-term interpretation of SC is adoption of NP. For example, NP brings potential long-term benefits to fisheries by spillover effects and larval dispersal. Roberts et al. (2005, pp. 123–124) asserted that 'fully protected marine reserves…can simultaneously meet conservation and fishery management objectives. Within their boundaries, they protect animals and their habitats; beyond their boundaries, they can enhance surrounding fisheries by emigration of animals and export of their offspring'. Indeed, Cullis-Suzuki and Pauly (2010, p. 114) argued that MRs are a subsidy for the fishing industry: 'the cost of maintaining MPAs (i.e., their running cost) can be considered a subsidy to fisheries…MPAs are…seen as "beneficial" subsidies, with any short-term loss by fishers outweighed by long-term, sustainable gains'. Alternatively, NP may morph into SC, for example when selected NTMPAs are seen as an element of SC strategy.

However, despite these five scenarios of possible *empirical* reconciliation between SC and NP, there remain fundamental *normative* differences between them. Although Rice et al. (2012, p. 218, 228) saw 'an increasing convergence of views on a number of key issues related to MPAs', they noted that 'For fisheries management the priorities are typically sustainable human use and food security while for environmental management the priorities are maintenance of biodiversity and ecosystem processes that underpin natural resource productivity.' Jones (2014) argued that any détente between NP and SC is too fragile to survive because the two sides are driven by diametrically opposed priorities—MSY and ecosystem restoration respectively. This difference reflects 'differing ethical perspectives, NTMPA proponents being more influenced by preservationist and ecocentric perspectives, and CFMA proponents being more influenced by the utilitarian resource conservation perspective' (Jones 2007, p. 38). According to Jones (2014, pp. 51–52, 53) this difference in priorities is a fundamental moral divide that is unbridgeable: 'These different value priorities…are the main reason why…it seems unlikely that the dispute between CFMA and no-take MPA proponents will disappear. as this dispute is underlain by different ethical perspectives, therefore 'consensus is philosophically intolerable'…scientific consensus on the need for extensive networks of no-take MPAs seems likely to remain elusive'. So Jones (2007, p. 38) is not hopeful for reconciliation between NP and SC: 'It seems likely that these opposing storylines will continue to be a major obstacle to further NTMPA designations'.

Our view is more optimistic. On the central question of whether the irreducible normative differences between NP and SC will prevent their respective advocates from engaging in discussions which explore the empirical routes of reconciliation through complementarity and compromise, our guess is that while hardliners on each side may be unwilling to contemplate such discussions, most NPs and SCs would rather reach a practical accommodation with the other side than risk the other side triumphing in full.

7.4 Conclusion

In this final chapter we have summarised the findings of previous chapters and spelled out some of the wider implications of these findings. The central messages we wish to convey in the book are fourfold. The first message is that the issue of MRs is hotly contested between those (whom we have named NPs) who see MRs as the crucial solution to the problems of over-exploitation of the world's marine fisheries resources, and those (whom we have named SCs) who see MRs as only one of a large number of measures to deal with overfishing. The second message is that the NPs, who in recent years have enjoyed a huge amount of support from both the scientific community and the broader environmental movement, have been accused of making more claims for MRs than the scientific evidence justifies. The third message is that there has been a reaction against NP from SCs, who have argued with some credibility that if CFM were to be fully enforced, it would deal more effectively than would MRs with the problem of overfishing, at less socio-economic cost to fishers. The fourth message is that, although there is a deep normative divide between the NP ideology (which extols the intrinsic value of biodiversity) and the SC ideology (which extols the instrumental value of biodiversity), there are practical ways in which their different empirical prescriptions about MRs may be reconciled. Taking such practical steps would depend on political will on each side, but the reward could be at least a partial win-win for everyone. This would be 'smart' MPA policymaking (Day 2008; Klein et al. 2008a).

[illegible]

7.4 Conclusion

In this final chapter we have summarized the [illegible]

References

Abbott J, Haynie A (2012) What are we protecting? Fisher behaviour and the unintended consequences of spatial closures as a fishery management tool. Ecol Appl 22(3):762–777
Abdulla A, Gomei M, Hyrenbach D, Notarbartolo-di-Sciara G, Agardy T (2009) Challenges facing a network of representative marine protected areas in the Mediterranean: prioritising the protection of underrepresented habitats. ICES J Marine Sci 66:22–28
Aburto-Oropeza O, Erisman B, Halland GR, Mascarenas-Osorio I, Sala E, Ezcura E (2011) Large recovery of fish biomass in a no-take marine reserve. PLoS ONE 6(8):1–7
Agardy T (2000) Effects of fisheries on marine ecosystems: a conservationist's perspective. ICES J Marine Sci 57(3):761–765
Agardy T (2010) Science and scarcity: is monitoring the first to go when economic times get tough? Marine Ecosyst Manage 4(3):7
Agardy T et al [10 co-authors] (2003) Dangerous targets? Unresolved issues and ideological clashes around marine protected areas. Aquat Conserv Marine Freshw Ecosyst 13(4):353–367
Agardy T, di Sciara GN, Christie P (2011) Mind the gap: addressing the shortcomings of marine protected areas through large scale marine spatial planning. Marine Policy 35(2):226–232
Airame S, Dugan JE, Lafferty KD, Leslie H, McArdle DA, Warner RR (2003) Applying ecological criteria to marine reserve design: a case study from the California Channel Islands. Ecol Appl 13(1):170–184
Alcala A, Russ GR (2006) No-take marine reserves and reef management in the Philippines: a new people power revolution. Ambio 35(5):245–254
Allison GW, Lubchenco J, Carr MH (1998) Marine reserves are necessary but not sufficient for marine conservation. Ecol Appl 8(1):79–92
AMSA [Australian Marine Sciences Association] (2008) NSW AMSA position statement on marine protected areas and no-take marine sanctuaries. AMSA, New South Wales
Andras P (2011) Research: metrics, quality, and management implications. Res Eval 20(2):90–106
Angulo-Valdes JA, Hatcher BG (2010) A new typology of benefits derived from marine protected areas. Marine Policy 34:635–644
Appleby T, Jones P (2012) The marine and coastal access act- A hornets' nest? Marine Policy 36:73–77
Ardron JA (2008a) The challenge of assessing whether the OSPAR network of marine protected areas is ecologically coherent. Hydrobiologia 606:45–53
Ardron JA (2008b) Three initial OSPAR tests of ecological coherence: heuristics in a data-limited situation. ICES J Marine Sci 65:1527–1533
Ardron K, Gjerde K, Pulle S, Tilot V (2008) Marine spatial planning in the high seas. Marine Policy 32:832–939
Ashworth JS, Ormond RFG (2005) Effects of fishing pressure and trophic group on abundance and spillover across boundaries of a no-take zone. Biol Conserv 121(3):333–344

A. Caveen et al., *The Controversy over Marine Protected Areas*,
SpringerBriefs in Environmental Science, DOI 10.1007/978-3-319-10957-2

Ashworth J, Aish A, Stoker B (2010) Marine conservation zone project: ecological network guidance. Natural England, Peterborough

Attwood CG, Bennett BA (1994) Variation in dispersal of Galjoen (*Coracinus capensis*) (Teleostei, Coracinidae) from a marine reserve. Can J Fish Aquat Sci 51(6):1247–1257

Babcock R, Shears NT, Alcala AC, Barrett NS, Edgar GJ, Lafferty KD, McClanahan TR, Russ GR (2010) Decadal trends in marine reserves reveal differential rates of change in direct and indirect effects. PNAS 107(43):18256–18261

Ball IR, Possingham HP (2000) MARXAN (V1.8.2): marine reserve design using spatially explicit annealing. manual. Manual prepared for the Great Barrier Reef Marine Park Authority

Ballantine W (1997) Design principles for systems of "notake" marine reserves. In: Pitcher TJ (ed) The design and monitoring of marine reserves. Vancouver Fisheries Centre Research Report vol. 5, University of British Columbia, Vancouver

Ballantine W (1999) Marine reserves in New Zealand: the development of the concept and principles. Proceedings of a workshop on marine conservation. Korean Ocean Research and Development Institute, Cheju Islands, Nov 1999:3–38

Ballantine B (2014) Fifty years on: lessons from marine reserves in New Zealand and principles for a worldwide network. Biol Conserv (in press)

Balmford A, Gravestock P, Hockley N, McLean CJ, Roberts CM (2004) The worldwide costs of marine protected areas. PNAS 101(26):9694–9697

Ban NC (2009) Minimum data requirements for designing a set of marine protected areas, using commonly available abiotic and biotic datasets. Biodivers Conserv 18(7):1829–1845

Ban NC, Vincent ACJ (2009) Beyond marine reserves: exploring the approach of selecting areas where fishing is permitted, rather than prohibited. PLoS ONE 4(7):e6528

Banks SA, Skilleter GA (2010) Implementing marine reserve networks: a comparison of approaches in New South Wales and New Zealand. Marine Policy 34:197–207

Banobi J, Branch T, Hilborn R (2011) Do rebuttals affect future science? Ecosphere 2(3):1–11

Barret N (1995) Short-term and long-term movement patterns of six temperate reef fishes (families Labridae and Monacanthidae). Marine Freshw Res 46(5):853–860

Barrett NS, Edgar GJ, Buxton CD, Haddon M (2007) Changes in fish assemblages following 10 years of protection in Tasmanian marine protected areas. J Exp Marine Biol Ecol 345(2):141–157

Barto E, Rillig M (2012) Dissemination biases in ecology: effect sizes matter more than quality. Oikos 121:228–235

Batagelj V, Mrvar A (2004) Pajek: package for large network analysis. Kosovo, University of Ljubljana

Baum JK, Worm B (2009) Cascading top-down effects of changing oceanic predator abundances. J Animal Ecol 78(4):699–714

Beare D et al [14 co-authors] (2013) Evaluating the effect of fishery closures: lessons from the Plaice Box. J Sea Res 84:49–60

Bene C, Tewfik A (2003) Biological evaluation of marine protected area: evidence of crowding effect on a protected population of queen conch in the Caribbean. Marine Ecology-Pubblicazioni Della Stazione Zoologica Di Napoli I 24(1):45–58

Bennett BA, Attwood CG (1991) Evidence for recovery of a surf-zone fish assemblage following the establishment of a marine reserve on the southern coast of South Africa. Marine Ecol Prog Ser 75(2–3):173–181

Berkes F (2009) Evolution of co-management: role of knowledge generation, bridging organizations and social learning. J Environ Manage 90:1692–1702

Beukers-Stewart BD, Vause BJ, Mosley MWJ, Rossetti HL, Brand AR (2005) Benefits of closed area protection for a population of scallops. Marine Ecol-Prog Ser 298:189–204

Bjorklund MI (1974) Achievements in marine conservation I. Marine parks. Environ Conserv 1(3):205–217 (Autumn)

Blanck A, Lamouroux N (2007) Large-scale intraspecific variation in life-history traits of European freshwater fish. J Biogeogr 34(5):862–875

Bloomfield HJ, Sweeting CJ, Mill AC, Stead SM, Polunin NVC (2012) No-trawl area impacts: perceptions, compliance and fish. Environ Conserv 39(3):1–11

Blount BG, Pitchon A (2007) An anthropological research protocol for marine protected areas: creating a niche in multidisciplinary cultural hierarchy. Hum Org 66(2):103–111

Blyth-Skyrme RE, Kaiser MJ, Hiddink JG, Edwards-Jones G, Hart PJB (2006) Conservation benefits of temperate marine protected areas: variation among fish species. Conserv Biol 20(3):811–820

Bohnsack JA, Causey B, Crosby MP, Griffis RB, Hixon MA, Hourigan TF, Koltes KH, Maragos JE, Simons A, Tilmant JT (2000) A rationale for minimum 20–30% no-take protection. Proceedings of the 9th International Coral Reef Symposium. Bali, 23–27 Oct 2000

Boncoeur J, Alban F, Thebaud O (2011) Bioeconomic analysis of marine protected area fisheries effects. In: Claudet J (ed) Marine protected areas: a multidisciplinary approach. Cambridge University Press, Cambrige

Botsford LW, Micheli F, Hastings A (2003) Principles for the design of marine reserves. Ecol Appl 13(1):25–31

Boyd I (2013) Point of view: making science count in government: statement made by the Chief Scientific Advisor at Defra. http://dx.doi.org/10.7554/eLife2013;2:e01061. Accessed 2 July 2013

Brander K (2010) Reconciling biodiversity conservation and marine capture fisheries production. Curr Opin Environ Sustain 2(5–6):416–421

Brussard P, Tull J (2007) Conservation biology and four types of advocacy. Conserv Biol 21(1): 21–24

Canadas A, Sagarminaga R, De Stephanis R, Urquiola E, Hammond PS (2005) Habitat preference modelling as a conservation tool: proposals for marine protected areas for cetaceans in southern Spanish waters. Aquat Conserv Marine Freshw Ecosyst 15(5):495–521

Carrington D (2013) England names 27 new marine conservation zones. The Guardian 21.11.13

Carson HS, Hentschel BT (2006) Estimating the dispersal potential of polychaete species in the Southern California Bight: implications for designing marine reserves. Marine Ecol Prog Ser 316:105–113

Caveen AJ, Sweeting CJ, Willis TJ, Polunin NVC (2012) Are the scientific foundations of temperate marine reserves too warm and hard? Environ Conserv 39(3):199–203

Caveen AJ, Gray TS, Stead SM, Polunin NVC (2013) MPA policy: what lies behind the science? Marine Policy 37(1):128–134

CBD (1994) Report of the first meeting on the conference of the parties to the Convention on Biological Diversity. Nassau, 28th Nov–9th Dec 1994

CEFAS [Centre for Environment, Fisheries and Aquaculture Science] (2005) Investigations into closed area management of the North Sea cod. CEFAS, Lowestoft

Charles A, Wilson L (2009) Human dimensions of marine protected areas. ICES J Marine Sci 66:6–15

Chateau O, Wantiez L (2009) Movement patterns of four coral reef fish species in a fragmented habitat in New Caledonia: implications for the design of marine protected area networks. ICES J Marine Sci 66(1):50–55

Chiappone M, Sealey KMS (2000) Marine reserve design criteria and measures of success: lessons learned from the Exuma Cays L and Sea Park, Bahamas. Bull Marine Sci 66(3):691–705

Christie P (2004) Marine protected areas as biological successes and social failures in Southeast Asia. Am Fish Soc Symp 42:155–164

Christie P (2011) Creating space for interdisciplinary marine and coastal research: five dilemmas and suggested resolutions. Environ Conserv 38(2):172–186

Christie MR, Tissot BN, Albins MA, Beets JP, Jia Y, Ortiz DM, Thompson SE, Hixon MA (2010) Larval connectivity in an effective network of marine protected areas. PLoS ONE 5(12):1–8

Cinner JE, Daw T, Huchery C, Thoya P, Wamukota A, Cedras M, Abunge C (forthcoming) Winners and Losers in marine conservation: fishers' displacement and livelihood benefits from marine reserves. Soc Nat Resour

Clark CW (1996) Marine reserves and the precautionary management of fisheries. Ecol Appl 6(2):369–370

Clarke J (2013) NWIFCA letter to Minister on MCZs. Fishing News 23.8.13:5

Claudet J, Pelletier D, Jouvenel JY, Bachet F, Galzin R (2006) Assessing the effects of marine protected area (MPA) on a reef fish assemblage in a Northwestern Mediterranean marine reserve: identifying community based indicators. Biol Conserv 130(3):349–369

Claudet J, Guidetti P, Mouillot D, Shears NT, Micheli F (2011) Ecological effects of marine protected areas: conservation, restoration and functioning. In: Claudet J (ed) Marine protected areas: a multidisciplinary approach. Cambridge University Press, Cambridge

Clover C (2004) The end of the line. Ebury Press, London

Cohen D, de la Vega R, Watson G (2001) Advocacy for social justice. Kumarian Press, West Hartford

Cole RG, Villouta E, Davidson RJ (2000) Direct evidence of limited dispersal of the reef fish *Parapercis colias* (Pinguipedidae) within a marine reserve and adjacent fished areas. Aquat Conserv Marine Freshw Ecosyst 10(6):421–436

Connell J (2013) Islands at risk: environments, economies and contemporary change. Edward Elgar, Cheltenham

Conover DO, Munch SB (2002) Sustaining fisheries yields over evolutionary time-scales. Science 297:94–96

Corrigan CM, Kershaw F (2008) Working toward high seas marine protected areas. UNEP-WCMC, Cambridge

Cǫté IM, Mosquera I, Reynolds JD (2001) Effects of marine reserve characteristics on the protection of fish populations: a meta-analysis. J Fish Biol 59(Supplement A):178–189

Cressey D (2011a) Troubled waters. Nature 480:151–152

Cressey D (2011b) Uncertain sanctuary. Nature 480:166–167

Cudney-Bueno R, Lavin MF. Marinone SG, Raimondi PT, Shaw WW (2009) Rapid effects of marine reserves via larval dispersal. PLoS ONE 4(1):7

Cullis-Suzuki S, Pauly D (2010) Marine protected area costs as 'beneficial' fisheries subsidies: a global evaluation. Coast Manage 38:113–121

Curley BG, Kingsford MJ, Gillanders BM (2002) Spatial and habitat-related patterns of temperate reef fish assemblages: implications for the design of marine protected areas. Marine Freshw Res 53(8):1197–1210

Cvitanovic, C et al [36 co-authors] (2013) Critical research needs for managing coral reef marine protected areas: perspectives of academics and managers. J Environ Manage 114:84–91

David GE (2005) Science and society: marine reserve design for the California Channel Islands. Conserv Biol 19(6):1745–1751

Davis G (1999) Why don't parks and sanctuaries protect marine fish too? The George Wright Forum 16(2):88–96

Daw T, Gray T (2005) Fisheries science and sustainability in international policy: a study of failure in the European Union's common fisheries policy. Marine Policy 29:189–197

Day J (2008) The need and practice of monitoring, evaluating and adapting marine planning and management—lessons from the Great Barrier Reef. Marine Policy 32(5):823–831

Dayton PK, Thrush SF, Agardy MT, Hofman RJ (1995) Environmental effects of marine fishing. Aquat Conserv Marine Freshw Ecosyst 5(3):205–232

De Nooy W, Mrvar A, Batagelj V (2005) Exploratory social network analysis with Pajek. Cambridge University Press, Cambridge

De Santo EM (2013) Missing marine protected area (MPA) targets: how the push for quantity over quality undermines sustainability and social justice. J Environ Manage 124:137–146

De Santo EM, Jones PJS, Miller AMM (2011) Fortress conservation at sea: a commentary on the Chagos marine protected area. Marine Policy 35(2):258–260

de Segura AG, Tomas J, Pedraza SN, Crespo EA, Rago, JA (2003) Preliminary patterns of distribution and abundance of loggerhead sea turtles, *Caretta caretta*, around Columbretes Islands Marine Reserve, Spanish Mediterranean. Marine Biol 143(4):817–823

De Silva ME, Gately EM, Desilvestre I (1986) A bibliographic listing of coastal and marine protected areas: a global survey. Woods Hole Oceanographic Institution, Massachusetts

Defra (2002) Safeguarding our seas: a strategy for the conservation and sustainable development of our marine environment. Department for Environment, Food, and Rural Affairs, London

Defra (2004) Review of marine nature conservation. Department for Environment, Food, and Rural Affairs, London

Defra (2006a) A Marine Bill. A consultation document by the Department for Environment, Food and Rural Affairs. Summary of responses. Department for Environment, Food and Rural Affairs, London

Defra (2006b) A Marine Bill: a consultation document. Department for Environment, Food and Rural Affairs, London

Defra (2006c) The UK government response to the Royal Commission on Environmental Pollution's twenty-fifth report. Department for Environment, Food, and Rural Affairs, London

Defra (2008a) Marine and Coastal Access Bill impact assessment. Department for Environment, Food and Rural Affairs, London

Defra (2008b) Draft guidance on selection and designation of marine conservation zones (note 1). Department for Environment, Food and Rural Affairs, London

Defra (2009) Marine and Coastal Access Act. Department for Environment, Food and Rural Affairs, London

Defra (2010) UK marine science strategy: shaping, supporting, co-ordinating and enabling the delivery of world class marine science for the UK. 2010–2025. Department for Environment, Food, and Rural Affairs, London

Degnbol P, Gislason H, Hanna S, Jentoft S, Raakjaer Nielson J, Sverdrup-Jensen S, Wilson DC (2006) Painting the floor with a hammer: technical fixes in fisheries management. Marine Policy 30(5):534–543

Denny CM, Babcock RC (2004) Do partial marine reserves protect reef fish assemblages? Biol Conserv 116(1):119–129

Devillers R, Pressey RL, Grech A, Kittinger JH, Edgar GJ, Ward T, Watson R (2014) Reinventing residual reserves in the sea: are we favouring ease of establishment over need for protection? Aquat Conserv Marine Freshw Ecosyst (in press)

Dinmore TA, Duplisea DE, Rackham BD, Maxwell DL, Jennings S (2003) Impact of a large-scale area closure on patterns of fishing disturbance and the consequences for benthic communities. ICES J Marine Sci 60(2):371–380

Dudley N (2008) Guidelines for applying protected area management categories. Gland, International Union for Conservation of Nature

Dulvy NK (2006) Conservation biology: strict marine protected areas prevent reef shark declines. Curr Biol 16(23):R989–R991

Dulvy NK (2013) Super-sized MPAs and the marginalization of species conservation. Aquat Conserv Marine Freshw Ecosyst 23:357–362

Dunne RP, Newman S, Papastavrou V, Polunin N, Sand PH (2014) The creation of the Chagos marine protected area: a fisheries perspective. (Forthcoming)

Eades S (2012) Consultation on MSFD: UK initial assessment and proposals for GES. MARINET (Marine renewable infrastructure network). Cork, University College

Eastwood P, Mills CM, Aldridge JN, Houghton CA, Rogers SI (2007) Human activities in UK offshore waters: an assessment of direct, physical pressure on the seabed. ICES J Marine Sci 64:453–463

EC [European Commission] (2002) Council regulation (EC) No 2371/2002 of 20 December 2002 on the conservation and sustainable exploitation of fisheries resources under the Common Fisheries Policy

Eden L (2009) Speech in the House of Lords on the Marine and Coastal Access Bill, Committee (6th Day), 12 March 2009, columns 708–709

Edgar G (2011) Does the global network of marine protected areas provide an adequate safety net for marine biodiversity? Aquat Conserv Marine Freshw Ecosyst 21:313–316

Edgar GJ, Stuart-Smith RD (2009) Ecological effects of marine protected areas on rocky reef communities-a continental-scale analysis. Marine Ecol-Prog Ser 388:51–62

Edgar GJ et al [24 co-authors] (2014) Global conservation outcomes depend on marine protected areas with five key features. Nature 506:216–228

Edgar GJ, Barrett NS, Stuart-Smith RD (2009) Exploited reefs protected from fishing transform over decades into conservation features otherwise absent from seascapes. Ecol Appl 19(8):1967–1974

Edwards CTT, Rademeyer RA, Butterworth DS, Plaganyi EE (2009) Investigating the consequences of marine protected areas for the South African deep-water hake (*Merluccius paradoxus*) resource. ICES J Marine Sci 66:72–81

Elliott M, Hargreaves J, Pilgrim S (2012) UK sea fisheries statistics 2011. Defra, London

Fanelli D (2012) Negative results are disappearing from most disciplines and countries. Scientometrics 90:891–904

Fanshawe S, Van Blaricom GR, Shelly AA (2003) Restored top carnivores as detriments to the performance of marine protected areas intended for fishery sustainability: a case study with red abalones and sea otters. Conserv Biol 17(1):273–283

FAO [Food and Agriculture Organization of the United Nations] (2006) The state of world highly migratory, Straddling and other high seas fishery resources and associated species. FAO fisheries technical paper 495. FAO, Rome

FAO (2009) Marine protected areas in the high seas. FAO, Rome

FAO (2012) The state of world fisheries and aquaculture. FAO, Rome

Farrow S (1996) Marine protected areas: emerging economics. Marine Policy 20(6):439–446

Fernandes L et al [15 co-authors] (2005) Establishing representative no-take areas in the Great Barrier Reef: large-scale implementation of theory on marine protected areas. Conserv Biol 19(6):1733–1744

Finlayson AC (1994) Fishing for truth: a sociological analysis of northern cod stock assessments from 1977–1990. Institute of Social and Economic Research, Memorial University of Newfoundland St John's, Canada

Fischer F (2003) Reframing public policy: discursive politics and deliberative practices. Oxford University Press, Oxford

Fisher JAD, Frank KT (2002) Changes in finfish community structure associated with an offshore fishery closed area on the Scotian Shelf. Marine Ecol-Prog Ser 240:249–265

Fleming DM, Jones PJS (2012) Challenges to achieving greater and fairer stakeholder involvement in marine spatial planning as illustrated by the Lyme Bay scallop dredging closure. Marine Policy 36(2):370–377

Fletcher S, Jefferson R, Glegg G, Rodwell L, Dodds W (2014) England's evolving marine and coastal governance framework. Marine Policy 45:261–268

Fock H (2011) Natura 2000 and the European Common Fisheries Policy. Marine Policy 35(2): 181–188

Fogarty M, Murawski S (1998) Large-scale disturbance and the structure of marine system: fishery impacts on Georges Bank. Ecol Appl 8(1):6–22

Foucault M (1980) Power/knowledge: selected interviews and other writings. Harvester, Brighton

Francini-Filho RB Moura RL (2008) Evidence of spillover of reef fishes from a no-take marine reserve: an evaluation using the before-after control-impact (BACI) approach. Fish Res 93(3):346–356

Frank KT, Shackell NL, Simon JE (2000) An evaluation of the Emerald/Western Bank juvenile haddock closed area. ICES J Marine Sci 57(4):1023–1034

Franks JS (2000) A review: pelagic fishes at petroleum platforms in the northern Gulf of Mexico; diversity, interrelationships and perspectives. In: Le Gall J-Y, Cayre P, Taquet M (eds) Peche thoniere et dispositifs de concentration de poisons. Ifremer Actes Colloq 28:502–515

Fraschetti S, Claudet J, Grorud-Colvert K (2011) Transitioning from single sector management to ecosystem-based management: what can marine protected areas offer? In: Claudet J (ed) Marine protected areas: a multidisciplinary approach. Cambridge University Press, Cambridge

Fraser HM, Greenstreet SPR, Piet GJ (2009) Selecting MPAs to conserve groundfish biodiversity: the consequences of failing to account for catchability in survey trawls. ICES J Marine Sci 66:82–89

Frid CLJ, Paramor O (2006) Marine biodiversity: the rationale for intervention. Report to Defra

Frid C, Paramor O, Scott C (2006) Ecosystem-based management of fisheries: is science limiting? ICES J Marine Sci 63(9):1567–1572

Fulton E, Smith ADM, Smith DC, van Putten IE (2011) Human behaviour: the key source of uncertainty in fisheries management. Fish Fisheries 12(1):2–17

Funtowicz S, Ravetz J (1991) A new scientific methodology for global environmental issues. In: Costanza R (ed) The ecological economics. Columbia University Press, New York

Gaines SD, Lester SE, Grorud-Colvert K, Costello C, Pollnac R (2010a) Evolving science of marine reserves: new developments and emerging research frontiers. PNAS 107(43): 18251–18255

Gaines SD, White C, Carr MH, Palumbi SR (2010b) Designing marine reserve networks for both conservation and fisheries management. PNAS 107(43):18286–18293

Galal N, Ormond RFG, Hassan O (2002) Effect of a network of no-take reserves in increasing catch per unit effort and stocks of exploited reef fish at Nabq, South Sinai. Marine Freshw Res 53(2):199–205

Game ET, Watts ME, Wooldridge S, Possingham HP (2008) Planning for persistence in marine reserves: a question of catastrophic importance. Ecol Appl 18(3):670–680

Gell FR, Roberts CM (2003) Benefits beyond boundaries: the fishery effects of marine reserves. Trends Ecol Evol 18(9):448–455

Gerber LR, Botsford LW, Hastings A, Possingham HP, Gaines SD, Palumbi SR, Andelman S (2003) Population models for marine reserve design: a retrospective and prospective synthesis. Ecol Appl 13(1):47–64

Gerber LR, Heppell SS, Ballantyne F, Sala E (2005) The role of dispersal and demography in determining the efficacy of marine reserves. Can J Fish Aquat Sci 62:863–871

Gerhardinger LC, Godoy EAS, Jones PJS, Sales G, Ferreira BP (2011) Marine protected dramas: the flaws of the Brazilian national system of marine protected areas. Environ Manage 47(4):630–643

Gladstone W (2007) Requirements for marine protected areas to conserve the biodiversity of rocky reef fishes. Aquat Conserv Marine Freshw Ecosyst 17:71–87

Gleason M et al [14 co-authors] (2013) Designing a network of marine protected areas in California: achievements, costs, lessons learned, and challenges ahead. Ocean Coast Manage 74:90–101

Glenn H et al [12 co-authors] (2012) Trust in the fisheries scientific community. Marine Policy 36(1):54–72

GOC [Global Ocean Commission] (2013) MPAs: protecting high seas biodiversity. Policy options paper #7. Papers prepared for the 3rd meeting of the GOC. Nov 2013. http://www.globalocean-commission.org/policies/. Accessed 4 Dec 2013

Godfray H, Beddington J, Crute I (2010) Food security: the challenge of feeding 9 billion people. Science 327:812–818

Goni R, Badalamenti F, Tupper MH (2011) Effects of marine protected areas on local fisheries: evidence from empirical studies. In: Claudet J (ed) Marine protected areas: a multidisciplinary approach. Cambridge University Press, Cambridge

Goodwin P (1999) The end of consensus? The impact of participatory initiatives on conceptions of conservation and the countryside in the United Kingdom. Environ Plan D 17(4):383–401

Grafton R, Akter S, Kompas T (2011) A policy-enabling framework for the ex-ante evaluation of marine protected areas. Ocean Coast Manage 54(6):478–487

Graham NAJ et al [15 co-authors] (2008) Climate warming, marine protected areas and the ocean-scale integrity of coral reef ecosystems. PLoS ONE 3(8):1–9

Graham NAJ et al [17 co-authors] (2011) From microbes to people: tractable benefits of no-take areas for coral reefs. Oceanogr Marine Biol Annu Rev 49:105–136

Gray M (2010) Letter from Seafish's environmental assessment officer to Dr Jen Ashworth. Natural England, 13.4.10

Gray N, Campbell L (2008) Science, policy advocacy, and marine protected areas. Conserv Biol 23(2):460–468

Gray T, Hatchard J (2008) A complicated relationship: stakeholder participation and the ecosystem-based approach to fisheries management. Marine Policy 32:158–168
Greenstreet SPR, Fraser HM, Piet GJ (2009) Using MPAs to address regional-scale ecological objectives in the North Sea: modelling the effects of fishing effort displacement. ICES J Marine Sci 66(1):90–100
Grorud-Colvert K, Claudet J, Carr M, Caselle J, Day J, Friedlander A, Lester SE, Lison de Loma T, Tissot B, Malone D (2011) The assessment of marine reserve networks: guidelines for ecological evaluation. In: Claudet J (ed) Marine protected areas: a multidisciplinary approach. Cambridge University Press, Cambridge
Grorud-Colvert K, Lester SE, Airame S, Neeley E, Gaines SD (2010) Communicating marine reserve science to diverse audiences. PNAS 107(43):18306–18311
Grossman G, Jones G, Seaman W (1997) Do artificial reefs increase regional fish production? A review of existing data. Fisheries 22(4):17–23
Guerreiro J, Chircop A, Grilo C, Viras A, Ribeiro R, van der Elst R (2010) Establishing a transboundary network of marine protected areas: diplomatic and management options for the east African context. Marine Policy 34:896–910
Guidetti P (2002) The importance of experimental design in detecting the effects of protection measures on fish in Mediterranean MPAs. Aquat Conserv Marine Freshw Ecosyst 12(6):619–634
Guidetti P, Claudet J (2010) Comanagement practices enhance fisheries in marine protected areas. Conserv Biol 24(1):312–318
Guidetti P et al [17 co-authors] (2008) Italian marine reserve effectiveness: does enforcement matter? Biol Conserv 141(3):699–709
Guidetti P, Bussotti S, Pizzolante F, Ciccolella A (2010) Assessing the potential of an artisanal fishing co-management in the marine protected area of Torre Guaceto (southern Adriatic Sea, SE Italy) Fish Res 101:180–187
Haas PM (1989) Do regimes matter? Epistemic communities and Mediterranean pollution-control. Int Organ 43(3):377–403
Haas PM (1991) Policy responses to stratospheric ozone depletion. Glob Environ Change-Hum Policy Dimens 1(3):224–234
Haas PM (1992) Epistemic communities and international-policy coordination- introduction. Int Organ 46(1):1–35
Habermas J (1984) The theory of communicative action; volume 1: reason and the rationalisation of society. Beacon Press, Boston
Hajer M (1995) The politics of environmental discourse: ecological modernisation and the policy process. Oxford University Press, Oxford
Hale LZ (2014) My main takeaways from IMPAC3. MPA News 15(4):5
Hall-Spencer JM, Moore PG (2000) Scallop dredging has profound, long-term impacts on maerl habitats. ICES J Marine Sci 57:1407–1415
Halpern BS (2003) The impact of marine reserves: do reserves work and does reserve size matter? Ecol Appl 13(1):117–137
Halpern BS, Warner RR (2002) Marine reserves have rapid and lasting effects. Ecol Letters 5(3):361–366
Halpern BS, Warner RR (2003) Matching marine reserve design to reserve objectives. Proc Royal Soc Lond B Biolog Sci 270:1871–1878
Halpern BS, Lester SE, McLeod KL (2010) Placing marine protected areas onto the ecosystem-based management seascape. PNAS 107(43):18312–18317
Hamilton SL, Caselle JE, Malone DP, Carr MH (2010) Incorporating biogeography into evaluations of the Channel Islands marine reserve network. PNAS 107(43):18272–18277
Hanna S (2003) The economics of protected areas in marine fisheries management: an overview of the issues. In: Shipley JB (ed) Symposium on aquatic protected areas as fisheries management tools. American Fisheries Society, Quebec City
Hannesson R (1998) Marine reserves: what would they accomplish? Marine Resour Econ 13: 159–70

Hart DR (2006) When do marine reserves increase fishery yield? Can J Fish Aquat Sci 63:1445–1449

Hastings A, Botsford LW (2003) Comparing designs of marine reserves for fisheries and for biodiversity. Ecol Appl 13(1):65–70

Hastings S (2009) A watery grave: why fish will fall off the menu unless we protect the seas. Yorkshire Post 23.5.09

Hattam CE, Mangi SC, Gall SC, Rodwell LD (2014) Social impacts of a temperate fisheries closure: understanding stakeholders' views. Marine Policy 45:269–278

Hawkins JP, Roberts CM, Dytham C, Schelten C, Nugues MM (2006) Effects of habitat characteristics and sedimentation on performance of marine reserves in St Lucia. Biol Conserv 127(4):487–499

Hedgecock D, Barber PH, Edmands S (2007) Genetic approaches to measuring connectivity. Oceanography 20:70–79

Hiddink JG, Hutton T, Jennings S, Kaiser MJ (2006) Predicting the effects of area closures and fishing effort restrictions on the production, biomass, and species richness of benthic invertebrate communities. ICES J Marine Sci 63(5):822–830

Hiddink JG, Jennings S, Kaiser MJ (2007) Assessing and predicting the relative ecological impacts of disturbance on habitats with different sensitivities. J Appl Ecol 44:405–413

Higgins RM, Vandeperre F, Perez-Ruzafa A, Santos RS (2008) Priorities for fisheries in marine protected area design and management: implications for artisanal-type fisheries as found in southern Europe. J Nat Conserv 16:222–233

Hilborn R (2002) Measuring the effects of marine reserves of fisheries: the dilemmas of experimental programs. MPA News 4:1–3

Hilborn R (2006) Faith-based fisheries. Fisheries 31(11):554–555

Hilborn R (2007a) Managing fisheries is managing people: what has been learned? Fish Fisheries 8:285–296

Hilborn R (2007b) Reinterpreting the state of fisheries and their management. Ecosystems 10(8):1362–1369

Hilborn R (2007c) Defining success in fisheries and conflicts in objectives. Marine Policy 13:153–158

Hilborn R (2007d) Moving to sustainability by learning from successful fisheries. Ambio 36(4):296–303

Hilborn R (2011) Future directions in ecosystem based fisheries management: a personal perspective. Fish Res 108(2–3):235–239

Hilborn R (2013) MPAs matching the tools to the objectives. Fish Res 144:111

Hilborn R, Ovando D (2014) Reflections on the success of traditional fisheries management. ICES J Marine Sci 71(5):1040–1046

Hilborn R et al [16 co-authors] (2004) When can marine reserves improve fisheries management? Ocean Coast Manage 47(3–4):197–205

Hiscock K (2013) Making progress with MCZs: stop digging and look for ladders. Unpublished personal view

Hiscock K (2014) MCZs and evidence. Conference paper at Coastal Futures 2014: review and future trends. 22–23 January 2014, SOAS, London

Hiscock K, Breckels M (2007) Marine biodiversity hotspots in the UK: a report identifying and protecting areas for marine biodiversity. WWF UK, Godalming

Hobday A, Game ET, Grantham HS, Richardson AJ (2011) Conserving the largest habitat on earth: protected areas in the pelagic ocean. In: Claudet J (ed) Marine protected areas: a multidisciplinary approach. Cambridge University Press, Cambridge

Hoefnagel E, Burnett A, Wilson D (2006) The knowledge base of co-management. In: Motos L, Wilson D (eds) The knowledge base for fisheries management. Pergamon Press, Oxford

Holland DS (2000) A bioeconomic model of marine sanctuaries on Georges Bank. Can J Fish Aquat Sci 57(6):1307–1319

Holland DS, Stokes TK (2006) Comment on 'Fishing and the impact of marine reserves in a variable environment'. Can J Fish Aquat Sci 63:1183–1185

Holland KN, Lowe CG, Wetherbee BM (1996) Movements and dispersal patterns of blue trevally (*Caranx melampygus*) in a fisheries conservation zone. Fish Res 25(3–4):279–292

Horwood JW, Nichols JH, Milligan S (1998) Evaluation of closed areas for fish stock conservation. J Appl Ecol 35(6):893–903

Hoskin MG, Coleman RA, von Carlshausen E, Davis CM (2010) Variable population responses by large decapod crustaceans to the establishment of a temperate marine no-take zone. Can J Fish Aquat Sci 68:185–200

Hughes TP, Bellwood DR, Folke CS, McCook LJ, Pandolfi JM (2006) No-take areas, herbivory and coral reef resilience. Trends Ecol Evol 22(1):1–3

Huntington BE (2011) Confronting publication bias in marine reserve meta-analyses. Front Ecol Environ 9(7):375–376

Hutchings K, Griffiths MH (2010) Life-history strategies of *Umbrina robinsoni* (Sciaenidae) in warm-temperate and subtropical South African marine reserves. Afr J Marine Sci 32(1):37–53

Irranca-Davies H (2009) Letter to Stephen Eades of MARINET. 11.11.09

IUCN (2008) Establishing resilient marine protected area networks—making it happen. International Union for Conservation of Nature, Gland

Jackson JBC et al [18 co-authors] (2001) Historical overfishing and the recent collapse of coastal ecosystems. Science 293(5530):629–638

Jameson SC, Tupper MH, Ridley JM (2002) The three screen doors: can marine 'protected' areas be effective? Marine Pollut Bull 44:1177–1183

Jamieson GS (2000) Marine protected areas and their relevance to abalone (Haliotis kamtschatkana) conservation in British Columbia. Can Special Publ Fish Aquat Sci 130:139–147

Jasanoff S (ed) (2006) States of knowledge: the co-production of science and the social order. Taylor and Francis, London

Jaworski A, Solmundsson J, Ragnarsson SA (2010) Fish assemblages inside and outside marine protected areas off northern Iceland: protection effects or environmental confounds? Fish Res 102(1–2):50–59

Jennings S (2009) The role of marine protected areas in environmental management. ICES J Marine Sci 66:16–21

Jennings S, Blanchard JL (2004) Fish abundance with no fishing: predictions based on macroecological theory. Animal Ecol 73(4):632–642

Jennings S, Kaiser M (1998) The effects of fishing on marine ecosystems. Adv Marine Biol 34:201–352

Jennings S, Lee J (2012) Defining fishing grounds with vessel monitoring system data. ICES J Marine Sci 69(1):51–63

Jensen E, Holliman R (2009) Investigating science communication to inform science outreach and public engagement. In:Holliman R, Whitelegg L, Scanlon E, Smidt S, Thomas S, Thomas J (eds) Investigating science communication in the information age: implications for public engagement and popular media. Oxford University Press, Oxford

Jentoft S, van Son TC, Bjorkan M (2007) Marine protected areas: a governance system analysis. Hum Ecol 35(5):611–622

JNCC [Joint Nature Conservation Committee] (2010) Additional guidance for regional MCZ projects on planning for areas where licensed, planned or existing socio-economic activities occur. Joint publication by Natural England and JNCC, Peterborough

Jones PJS (2002) Marine protected area strategies: issues, divergences and the search for middle ground. Rev Fish Biol Fish 11(3):197–216

Jones PJS (2006) Collective action problems posed by no-take zones. Marine Policy 30(2): 143–156

Jones PJS (2007) Point-of-view: arguments for conventional fisheries management and against no-take marine protected areas: only half of the story? Rev Fish Biol Fish 17(1):31–43

Jones PJS (2008) Fishing industry and related perspectives on the issues raised by no-take marine protected area proposals. Marine Policy 32(4):749–758

Jones PJS (2009) Equity, justice and power issues raised by no-take marine protected area proposals. Marine Policy 33:759–765

Jones N (2011) Marine protection goes large. Nature News 16 May 2011. doi:10.1038/news.2011.292

Jones PJS (2012) Marine protected areas in the UK: challenges in combining top-down and bottom-up approaches to governance. Environ Conserv 39(3):248–258

Jones PJS (2014) Governing marine protected areas: resilience through diversity. Routledge, London

Jones PJS, Burgess J (2005) Building partnership capacity for the collaborative management of marine protected areas in the UK: a preliminary analysis. J Environ Manage 77(3):227–243

Jones PJS, Carpenter A (2009) Crossing the divide: the challenges of designing an ecologically coherent and representative network of MPAs for the UK. Marine Policy 33(5):737–743

Jones PJS, De Santo EM, Qiu W, Vestergaad O (2013) Introduction: an empirical framework for deconstructing the realities of governing marine protected areas. Marine Policy 41:1–4

Kaiser MJ (2004) Marine protected areas: the importance of being earnest. Aquat Conserv Marine Freshw Ecosyst 14(6):635–638

Kaiser MJ (2005) Are marine protected areas a red herring or fisheries panacea? Can J Fish Aquat Sci 62(5):1194–1199

Kaiser MJ (2009) Letter to Seafish 11.2.09

Kareiva P (2006) Conservation biology: beyond marine protected areas. Curr Biol 16(14): R533–R535

Katsanevakis S et al [31 co-authors] (2011) Ecosystem-based marine spatial management: review of concepts, policies, tools, and critical issues. Ocean Coast Manage 54(11):807–820

Keck M, Sikkink K (1998) Activists beyond borders: advocacy networks in international politics. Cornell University Press, Ithaca

Kelleher G (1999) Guidelines for marine protected areas. IUCN World Commission on Protected Areas, Gland

Kelleher G, Kenchington R (1991) Guidelines for establishing marine protected areas. International Union for Conservation of Nature, Gland

Kelleher G, Bleakley C, Wells S (1995) A global representative system of marine protected area. International Union for Conservation of Nature, Gland

Kim J-E (2013) The incongruity between the ecosystem approach to high seas marine protected areas and the existing high seas conservation regime. Aegean Rev Law Sea 2:1–36

Kirlin J, Caldwell M, Gleason M, Weber M, Ugoretz J, Fox E, Miller-Henson M (2013) California's marine life protection act initiative: supporting implementation of legislation establishing a statewide network of marine protected areas. Ocean Coast Manage 74:3–13

Kjaersgaard J., Frost H (2008) Effort allocation and marine protected areas: is the North Sea Plaice Box a management compromise? ICES J Marine Sci 65(7):1203–1215

Klein CJ, Chan A, Kircher L, Cundiff AJ, Gardner N, Hrovat Y, Scholz A, Kendall BE, Airame S (2008a) Striking a balance between biodiversity conservation and socioeconomic viability in the design of marine protected areas. Conserv Biol 22(3):691–700

Klein C, Steinback C, Scholz AJ, Possingham HP (2008b) Effectiveness of marine reserve networks in representing biodiversity and minimizing impact to fisherman: a comparison of two approaches used in California. Conserv Lett 1(1):44–51

Koldewey HJ, Curnick D, Harding S, Harrison LR, Gollock M (2010) Potential benefits to fisheries and biodiversity of the Chagos Archipelago/British Indian Ocean Territory as a no-take marine reserve. Marine Pollut Bull 60:1906–1915

Koricheva J (2003) Non-significant results in ecology: a burden or a blessing in disguise? Oikos 102(2):397–401

Krasner SD (1983) International regimes. Cornell University Press, Ithaca, New York

Kuhn TS (1970) The structure of scientific revolutions, 2nd edn. University of Chicago Press, Chicago

Lackey R (2001) Values, policy, and ecosystem health. Bioscience 51(6):437–443

Lackey R (2007) Science, scientists, and policy advocacy. Conserv Biol 21(1):12–17

Ladle R, Jepson P, Whittaker R (2005) Scientists and the media: the struggle for legitimacy in climate change and conservation science. Interdiscip Sci Rev 30(3):231–240

Laffoley D (2012) Transcript of interview with Dan Laffoley, Vice-Chair of the World Commission on MPAs. 12.3.12

Laffoley D, Maltby E, Vincent MA, Mee L, Dunn E, Gilliland P, Hamer JP, Mortimer D, Pound D (2004) The ecosystem approach: coherent actions for marine and coastal environments. A report to the UK government. English Nature, Peterborough

Lauck T, Clark CW, Mangel M, Munro GR (1998) Implementing the precautionary principle in fisheries management through marine reserves. Ecol Appl 8(1):72–78

Lawton JH (2007) Ecology, politics and policy. J Appl Ecol 44(3):465–474

Le Quesne WJF (2009) Are flawed MPAs any good or just a new way of making old mistakes? ICES J Marine Sci 66:132–136

Leake J (2012) Only 100 adult cod in North Sea. The Sunday Times 16.9.12

Lean G (2009) Save our seas: ministers go back on promises to protect UK waters. The Independent 8.2.09

Leisher C, van Beukering P, Scherl LM (2007) Nature's investment bank: how marine protected areas contribute to poverty reduction. The Nature Conservancy, Arlington

Leisher C et al [10 co-authors] (2012) Measuring the benefits and costs of community education and outreach in marine protected areas. Marine Policy 36(5):1005–1011

Lester SE, Halpern BS (2008) Biological responses in marine no-take reserves versus partially protected areas. Marine Ecol-Prog Ser 367:49–56

Lester SE, Halpern BS, Grorud-Colvert K, Lubchenco J, Ruttenberg BI, Gaines SD, Airame S, Warner RR (2009) Biological effects within no-take marine reserves: a global synthesis. Marine Ecol Prog Ser 384:33–46

Lincoln-Smith MP, Pitt KA, Bell JD, Mapstone BD (2006) Using impact assessment methods to determine the effects of a marine reserve on abundances and sizes of valuable tropical invertebrates. Can J Fish Aquat Sci 63(6):1251–1266

Link (2011) Marine Act policy bulletin—delivering a network of MPAs: highly protected sites. Wildlife and Countryside Link, London (June 2011)

Little LR, Grafton RQ, Kompas T, Smith AD, Punt AE, Mapstone BD (2010) Complementarity of no-take marine reserves and individual transferable catch quotas for managing the line fishery of the Great Barrier Reef. Conserv Biol 25(2):333–340

Lockwood S (2013a) Letter to Nigel Gooding, Defra 1.7.13

Lockwood S (2013b) 'Banal' Richard Benyon. The Times 5.8.13

Lockwood S (2013c) MPA Fishing Coalition counters Hugh. Fishing News 29.3.13:5

Louzao M, Hyrenbach DK, Arcos JM, Abello P, de Sola LG, Oro D (2006) Oceanographic habitat of an endangered Mediterranean procellariiform: implications for marine protected areas. Ecol Appl 16(5):1683–1695

Lubchenco J, Palumbi SR, Gaines SD, Andelman S (2003) Plugging a hole in the ocean: the emerging science of marine reserves. Ecol Appl 13(1):3–7

Lubchenco J, Gaines S, Grorud-Calvert K, Airame S, Palumbi S, Warner R, Smith BS (2007) The science of marine reserves, 2nd edn. PISCO. www.piscoweb.org

Lundquist CJ, Granek EF (2005) Strategies for successful marine conservation: integrating socio-economic, political, and scientific factors. Conserv Biol 19(6):1771–1778

MacGarvin M, Jones S (2000) Choose or lose: a recovery plan for fish stocks and the UK fishing industry. WWF UK, Surrey

MacMullen P (2009) Letter to The Independent on Sunday 8.2.09

MacMullen P (2010) Letter to Natural England 26.2.10

Maiorano L et al (13 co-authors] (2009) Systematic conservation planning in the Mediterranean: a flexible tool for the identification of no-take marine protected areas. ICES J Marine Sci 66(1):137–146

Mangi S, Hooper T, Rodwell L, Simon D, Snoxell D, Spalding M, Williamson P (eds) (2010) Establishing a marine protected area in the Chagos Archipelago: socio-economic considerations. Report of workshop, Royal Holloway, London University, London 7.1.10

Mangi S, Rodwell L, Hattam C (2011) Assessing the impacts of establishing MPAs on fishermen and fish merchants: the case of Lyme Bay, UK. Ambio 40:457–468

Margules C, Pressey R (2000) Systematic conservation planning. Nature 405:243–253
Marinesque S, Kaplan DM, Rodwell LD (2012) Global implementation of marine protected areas: is the developing world being left behind? Marine Politics 36:727–737
Marinet (2009) Amendments to the Marine and Coastal Access Bill. http://www.marinet.org.uk/campaign-article/marinet-amendments-to-the-uk-marine-bill-at-report-stage-in-the-house-of-lords-april-2009-2. Accessed 20 Aug 2009
Marris E (2006) Should conservation biologists push policies? Nature 442(7098):13
Marx W, Bornmann L (2013) The emergence of plate tectonics and the Kuhnian model of paradigm shift: a bibliometric case study based on the Anna Karenina principle. Scientometrics 94:595–614
Matz-Luck N, Fuchs J (2014) The impact of OSPAR on protected area management beyond national jurisdiction: effective regional cooperation or a network of paper parks? Marine Policy [In Press]
Mazur A (1981) The dynamics of technical controversy. Communications Press, Washington DC
McClanahan T (2011) Human and coral reef use interactions: from impacts to solutions? J Exp Marine Biol Ecol 408(1–2):3–10
McClanahan TR, Mangi S (2000) Spillover of exploitable fishes from a marine park and its effect on the adjacent fishery. Ecol Appl 10(6):1792–1805
McClanahan TR, Marnane J, Cinner JE, Kiene, WE (2006) A comparison of marine protected areas and alternative approaches to coral-reef management. Curr Biol 16:1408–1413
McClanahan TR et al [10 co-authors] (2008) Conservation action in a changing climate. Conserv Lett 1:53–59
McCook L et al [20 co-authors] (2010) Adaptive management of the Great Barrier Reef: a globally significant demonstration of the benefits of networks of marine reserves. PNAS 107(43): 18278–18285
McKie R (2013) Scientists accuse ministers of putting Britain's seas and marine life in peril. The Observer 4.8.13
McVittie A, Moran D (2010) Valuing the non-use benefits of marine conservation zones: an application to the UK Marine Bill. Ecol Econ 70:413–424
Merton RK (1973) The sociology of science: theoretical and empirical investigations. University of Chicago Press, Chicago
Metcalfe K (2013) Investigating the biological and socio-economic impacts of marine protected area network design in Europe. Unpublished PhD thesis, University of Kent, UK
Metcalfe K, Delavenne J, Garcia C, Foveau A, Dauvin J-C, Coggan R, Vaz S, Harrop SR, Smith R (2013) Impacts of data quality on the setting of conservation planning targets using the species-area relationship. Divers Distrib 19:1–13
Meyer CG, Holland KN, Wetherbee BM, Lowe CG (2000) Movement patterns, habitat utilization, home range size and site fidelity of whitesaddle goatfish, Parupeneus porphyreus, in a marine reserve. Environ Biol Fish 59(3):235–242
Micheli F, Halpern BS, Botsford JW, Warner RR (2004) Trajectories and correlates of community change in no-take marine reserves. Ecol Appl 14(6):1709–1723
Miller KI, Russ GR (2014) Studies of no-take marine reserves: methods for differentiating reserve and habitat effects. Ocean Coast Manage 96:51–60
Miller TR, Minteer BA, Malan L-C (2011) The new conservation debate: the view from practical ethics. Biol Conserv 144(3):948–957
Molenaar EJ, Oude Elferink, AG (2009) Marine protected areas in areas beyond national jurisdiction: the pioneering efforts under the OSPAR Convention. Utrecht Law Rev 5(1):5–20
Monbiot G (2012) The UK's marine reserves are nothing but paper parks. The Guardian 10.5.12
Mora C (2011) Effectiveness of the global network of marine protected areas. In: Claudet J (ed) Marine protected areas: a multidisciplinary approach. Cambridge University Press, Cambridge
Mora C, Andrefouet S, Costello MJ, Kranenburg C, Rollo A, Veron J, Gaston KJ, Myers RA (2006) Coral reefs and the global network of marine protected areas. Science 312:1750
Mosquera I, Côté IM, Jennings S, Reynolds J (2000) Conservation benefits of marine reserves for fish populations. Animal Conserv 4:321–332

Mouillot D, Culioli JM, Lepretre A, Tomasini JA (1999) Dispersion statistics and sample size estimates for three fish species (*Symphodus ocellatus, Serranus scriba* and *Diplodus annularis*) in the Lavezzi Islands Marine Reserve (South Corsica, Mediterranean Sea). Marine Ecology-Pubblicazioni Della Stazione Zoologica Di Napoli I 20(1):19–34

MPAC [Marine Protected Area Fishing Coalition] (nd) Coalition challenges scientific and legal basis of 'no take zones'. NFFO, York

Murawski SA, Brown R, Lai H-L, Rago PJ, Hendrickson L (2000) Large-scale closed areas as a fishery-management tool in temperate marine systems: the Georges Bank experience. Bull Marine Sci 66(3):775–798

Murawski SA, Wigley SE, Fogarty MJ, Rago PJ, Mountain DG (2005) Effort distribution and catch patterns adjacent to temperate MPAs. ICES J Marine Sci 62(6):1150–1167

Murtaugh P (2002) Journal quality, effect size, and publication bias in meta-analysis. Ecology 83(4):1162–1166

Myers RA, Worm B (2003) Rapid worldwide depletion of predatory fish communities. Nature 423(6937):280–283

Nardi K, Jones GP, Moran MJ, Cheng YW (2004) Contrasting effects of marine protected areas on the abundance of two exploited reef fishes of the sub-tropical Houtman Abrolhos Islands, Western Australia. Environ Conserv 31(2):160–168

NCEAS [National Center for Ecological Analysis and Synthesis] (2001) Scientific consensus statement on marine reserves and marine protected areas. Annual Meeting for the American Association for the Advancement of the Sciences, 17.2.01

Nelson M, Vucetich J (2009) On advocacy by environmental scientists: what, whether, why, and how. Conserv Biol 23(5):1090–1101

NFFO (2009a) Fisheries: the missing layer in marine planning. http://www.nffo.org.uk/news/fisheries_missing_layer.html. Accessed 11 Nov 2009

NFFO (2009b) The dangers of a 2.5 year marine protected area 'quick-fix' http://www.nffo.org.uk/news/dangers_of_marine_protected_area.html. Accessed 7 April 2009

NFFO (2010) Displacement from fishing grounds. http://www.nffo.org.uk/news/displacement_fishing.html. Accessed 13 Oct 2010

NFFO (2011) Fish robbers and crass journalism. http://www.nffo.org.uk/news/fish_robbers.html. Accessed 21 July 2011

NFFO (2012a) Coalition proposes new approach to MPAs. http://www.nffo.org.uk/news/Coalition_mpa.html. Accessed 25 Jan 2012

NFFO (2012b) MCZs: the opposite of the Big Society. http://www.nffo.org.uk/news/mczs.html. Accessed 25 Feb 2012

NFFO (2012c) The evidence-free campaign to limit fishing grounds. http://www.nffo.org.uk/news/fishingground_limit_2012.html. Accessed 14 Aug 2012

NFFO (nd) Marine and Coastal Access Bill: briefing note: achieving the right balance between conservation and socio-economic considerations. NFFO, York

NOAA [National Oceanographic and Atmospheric Administration] (2013) Conserving our oceans: one place at a time. National Marine Protected Areas Center, NOAA, Washington, DC

Nowotny H, Scott P, Gibbons M (2001) Re-thinking science: knowledge and the public in an age of uncertainty. Polity Press, Cambridge

Nowotny H, Scott P, Gibbons M (2003) 'Mode 2' revisited: the new production of knowledge. Minerva 41:179–194

O'Leary BC, Brown RL, Johnson DE, von Nordheim H, Ardron J, Packeiser T, Roberts CM (2012) The first network of marine protected areas (MPAs) in the high seas: the process, the challenges and where next. Marine Policy 36:598–605

Olsen EM et al [10 co-authors] (2013) Achieving ecologically coherent MPA networks in Europe: science needs and priorities. Marine Board position paper 18. Larkin KE, McDonough N (eds) European Marine Board, Ostend

O'Sullivan D, Emmerson M (2011) Marine reserve designation, trophic cascades and altered community dynamics. Marine Ecol Prog Ser 440:115–125

Osenberg CW, Shima JS, Miller SJ, Stier AC (2011) Assessing effects of marine protected areas: confounding in space and possible solutions. In: Claudet J (ed) Marine protected areas: a multidisciplinary approach. Cambridge University Press, Cambridge
OSPAR (2003a) Guidelines for the identification and selection of marine protected areas in the OSPAR maritime area. OSPAR Secretariat, London
OSPAR (2003b) Recommendation 2003/3 on a network of marine protected areas. OSPAR Secretariat, London
OSPAR (2010) OSPAR Recommendation 2010/2 on amending Recommendation 2003/3 on a network of Marine Protected Areas. OSPAR Secretariat, London
OSPAR (2013) 2012 Status report on the OSPAR network of marine protected areas. 618/2013. OSPAR Secretariat, London
Pajaro MG, Mulrennan ME, Vincent ACJ (2010) Towards an integrated marine protected areas policy: connecting the global to the local. Environ Dev Sustain 12:945–965
Parry S (2009) Stem cell scientists'discursive strategies for cognitive authority. Science as Culture 18(1):89–114
Pascoe S, Mardle S (2005) Economic impact of area closures and effort reduction measures in the North Sea. Report to Defra, London
Pastoors MA, Rijnsdorp AD, Van Beek FA (2000) Effects of a partially closed area in the North Sea ('plaice box') on stock development of plaice. ICES J Marine Sci 57(4):1014–1022
Pauly D, Froese R (2012) Comments on FAO's state of fisheries and aquaculture, or 'SOFIA 2010'. Marine Policy 36:746–752
Pauly D, Christensen V, Dalsgaard J, Froese R, Torres F (1998) Fishing down marine food webs. Science 279(5352):860–863
Pauly D, Christensen V, Guenette S, Pitcher TJ, Sumaila RU, Walters CJ, Watson R, Zeller D (2002) Towards sustainability in world fisheries. Nature 418(6898):689–695
Peckett FJ, Clegg GA, Rodwell LD (2014) Assessing the quality of data required to identify effective marine protected areas. Marine Policy 45:333–341
Pelc RA, Warner RR, Gaines SD, Paris CB (2010) Detecting larval export from marine reserves. PNAS 107(43):18266–18271
Peterson C, Lipcius R (2003) Conceptual progress towards predicting quantitative ecosystem benefits of ecological restorations. Marine Ecol Prog Ser 264:297–307
Phillipson J (2002) Widening the net: prospects for fisheries co-management. Centre for Rural Ecology', Newcastle University, Newcastle on Tyne, UK
Pickerell T (2013) Seafish: MCZ statement. http://www.seafish.org/about-seafish/blog/2013/11/21/seafish-mcz-statement. Accessed 25 Nov 2013
Pielke R (2004) When scientists politicize science: making sense of controversy over The Skeptical Environmentalist. Environ Sci Policy 7:405–417
Pielke RA (2007) The honest broker: making sense of science in policy and politics. Cambridge University Press, Cambridge
Pieraccini M (2013) Establishing an ecologically coherent network of marine protected areas in English waters: what does the designation of marine conservation zones under the Marine and Coastal Access Act 2009 add to the picture? Environ Law Rev 15:104–120
Pieraccini M (2014) (pers comm)
Pinnegar JK et al [10 co-authors] (2000) Trophic cascades in benthic marine ecosystems: lessons for fisheries and protected-area management. Environ Conserv 27(2):179–200
PISCO [Partnership for interdisciplinary studies of coastal oceans] (2007) The Science of marine reserves (Worldwide Edition). PISCO, California
PISCO (2011) The Science of marine reserves, (2nd Edn, Europe). PISCO, California
Pitcher TJ, Lam ME (2010) Fishful thinking: rhetoric, reality, and the sea before us: simple solutions to a complex problem. Ecol Soc 15(2):12
Pitchford JW, Codling EA, Psarra D (2007) Uncertainty and sustainability in fisheries and the benefit of marine protected areas. Ecol Model 207:286–292

Planes S (2011) Spacing a network of marine protected areas based on connectivity data. In: Claudet J (ed) Marine protected areas: a multidisciplinary approach. Cambridge University Press, Cambridge
PMSU [Prime Minister's Strategy Unit] (2004) Net benefits. PMSU, London
Polacheck T (2006) Tuna longline catch rates in the Indian Ocean: did industrial fishing result in a 90% rapid decline in the abundance of large predatory species? Marine Policy 30:470–482
Polunin NVC, Bloomfield HJ, Sweeting CJ, McCandless DT (2009) Effects of small prohibited trawling areas on the abundance of fishes: final report to the Esmée Fairbairn Foundation, April 2009
Pomeroy RS, Parks JE, Watson LM (2004) How is your MPA doing? A guidebook of natural and social indicators for evaluating marine protected area management effectiveness. IUCN, Gland
Pomeroy RS, Watson LM, Parks JE, Cid GA (2005) How is your MPA doing? A methodology for evaluating the management effectiveness of marine protected areas. Ocean Coast Manage 48:485–502
Potts T, Burdon D, Jackson E, Atkins J, Saunders J, Hastings E, Langmead O (2014) Do marine protected areas deliver flows of ecosystem services to support human welfare? Marine Policy 44:139–148
Pressey R (2013) Australia's new marine protected areas: why they won't work. The Conversation http://theconversation.com/australias-new-marine-protected-areas-why-they-wont-work-11469. Accessed 17 Jan 2013
Rakitin A, Kramer DL (1996) Effect of a marine reserve on the distribution of coral reef fishes in Barbados. Marine Ecol-Prog Ser 131(1–3):97–113
Rassweiler A, Costello C, Siegel DA (2012) Marine protected areas and the value of spatially optimized fishery management. PNAS 109(29):11884–11889
Ray GC (2004) Reconsidering 'dangerous targets' for marine protected areas. Aquat Conserv Marine Freshw Ecosyst 14(2):211–215
RCEP (2004) Turning the tide: addressing the impact of fisheries on the marine environment. Report to Defra, London
Rees SE, Rodwell LD, Atrill MJ, Austen MC, Mangi SC (2010) Is there a win–win scenario for marine nature conservation? A case study of Lyme Bay, England. Ocean Coast Manage 53(3):135–145
Rees SE, Attrill MJ, Austen MC, Mangi SC, Rodwell LD (2013) A thematic cost-benefit analysis of a marine protected area. J Environ Manage 114:476–485
Rees SE, Fletcher S, Gall SC, Friedrich LA, Jackson EL, Rodwell LD (2014) Securing the benefits: linking ecology with marine planning policy to examine the potential of a network of marine protected areas to support human wellbeing. Marine Policy 44:335–341
Ribeiro MC (2010) The 'Rainbow': the first national marine protected area proposed under the high seas. Int J Marine Coast Law 25:183–207
Rice JC (2011) Advocacy science and fisheries decision-making. ICES J Marine Sci 68(10): 2007–2012
Rice J, Houston K (2011) Representativity and networks of marine protected areas. Aquat Conserv Marine Freshw Ecosyst 21:649–657
Rice J et al [17 co-authors] (2012) The role of MPAs in reconciling fisheries management with conservation of biological diversity. Ocean Coast Manage 69:217–230
Richardson EA, Kaiser MJ, Hiddink JG, Galanidi M, Donald EJ (2006a) Developing scenarios for a network of marine protected areas: building the evidence base for the Marine Bill. Final report to Defra, 5.6.06
Richardson EA, Kaiser MJ, Edward-Jones G, Possingham HP (2006b) Sensitivity of marine-reserve design to the spatial resolution of socio-economic data. Conserv Biol 20(4):1191–1202
Rife AN, Erisman B, Sanchez A, Aburto-Oropeza O (2013) When good intentions are not enough… Insights on networks of 'paper park' marine protected areas. Conserv Lett 6(3):200–212
Robert CM, Hawkins JP, Gell FR (2005) The role of marine reserves in achieving sustainable fisheries. Phil Trans R Soc 360:123–132
Roberts CM (1997) Ecological advice for the global fisheries crisis. Trends Ecol Evol 12(1):35–38

Roberts CM (2000) Selecting marine reserve locations: optimality versus opportunism. Bulletin of Marine Sci 66(3):581–592

Roberts C (2003) Our shifting perspectives on the oceans. Oryx 37(2):166–177

Roberts CM (2007a) European scientists' consensus statement on marine reserves. York University, UK

Roberts CM (2007b) The unnatural history of the sea. Gaia, London

Roberts C (2012) Ocean of life. Penguin Books, London

Roberts C (nd) Critique of Kaiser (2005) Are marine protected areas a red herring or fisheries panacea? Unpublished paper

Roberts C, Hawkins J (1999) Extinction risk in the sea. Trends Ecol Evol 14(6):241–246

Roberts CM, Hawkins JP (2000) Fully-protected marine reserves: a guide. WWF Endangered Seas Campaign, Godalming, Surrey

Roberts C, Hilborn R (2013) Online debate: large no-take areas—is their total environmental impact positive or negative? A debate held on 8.10.13, sponsored by OpenChannels.org, MPA News, and the EBM Tools Network: 1–7

Roberts C, Mason L (2008) Return to abundance: a case for marine reserves in the North Sea. WWF UK, Godalming

Roberts CM, Bohnsack JA, Gell F, Hawkins JP, Goodridge R (2001) Effects of marine reserves on adjacent fisheries. Science 294(5548):1920–1923

Roberts CM et al [13 co-authors] (2003a) Ecological criteria for evaluating candidate sites for marine reserves. Ecol Appl 13(1):199–214

Roberts CM et al [11 co-authors] (2003b) Application of ecological criteria in selecting marine reserves and developing reserve networks. Ecol Appl 13(1):215–228

Robertson D, Hull R (2003) Public ecology: an environmental science and policy for global society. Environ Sci Policy 6(5):399–410

Rodmell D (2009) Correspondence with Defra official. 11.2.09

Rodmell D (2011) Marine conservation zone decisions too rushed: NFFO. Fish Information and Services.19.2.11

Rodwell LD, Roberts CM (2004) Fishing and the impact of marine reserves in a variable environment. Can J Fish Aquat Sci 61:2053–2068

Rodwell L et al [17 co-authors] (2014) Marine and coastal policy in the UK: challenges and opportunities in a new era. Marine Policy 45:251–258

Roff JC (2014) Networks of marine protected areas—the demonstrability dilemma. Aquat Conserv Marine Freshw Ecosyst 24:1–4

Rogers SI (1997) A review of closed areas in the United Kingdom exclusive economic zone. Science series technical report No106. CEFAS, Lowestoft

Rogers AD, Laffoley D, Polunin N, Tittensor DP (2013) Ocean conservation: current challenges and future opportunities. In: Macdonald DW, Willis KJ (eds) Key topics in conservation biology 2. Wiley, London

Rogers-Bennett L, Pearse JS (2001) Indirect benefits of marine protected areas for juvenile abalone. Conserv Biol 15(3):642–647

Rosenthal R (1979) The 'file drawer problem' and tolerance for null results. Psychol Bull 86:638–641

Rotherham D, Underwood AJ, Chapman MG, Gray CA (2007) A strategy for developing scientific sampling tools for fishery-independent surveys of estuarine fish in New South Wales, Australia. ICES J Marine Sci 64:1512–1516

Rowe S (2002) Population parameters of American lobster inside and outside no-take reserves in Bonavista Bay, Newfoundland. Fish Res 56(2):167–175

Royal Society (2006) Survey of factors affecting science communication by scientists and engineers. Royal Society, London

Rudd M, Tupper MH, Folmer H, van Kooten GC (2003) Policy analysis for tropical marine reserves: challenges and directions. Fish Fish 4:65–85

Rudershausen PJ, Mitchell WA, Buckel JA, Williams EH, Hazen E (2010) Developing a two-step fishery-independent design to estimate the relative abundance of deepwater reef fish: application to a marine protected area off the southeastern United States coast. Fish Res 105:254–260

Russ GR (2002) Yet another review of marine reserves as reef fishery management tools. In: Sale PF (ed) Coral reef fishes: dynamics and diversity in a complex ecosystem. Academic Press, Massachusetts

Russ GR, Alcala AC (1989) Effects of intense fishing pressure on an assemblage of coral reef fishes. Marine Ecol Prog Ser 56(1–2):13–27

Russ GR, Alcala AC (1996) Do marine reserves export adult fish biomass? Evidence from Apo Island, central Philippines. Marine Ecol-Prog Ser 132:1–9

Russ GR, Alcala AC (1998) Natural fishing experiments in marine reserves 1983–1993: roles of life history and fishing intensity in family responses. Coral Reefs 17(4):399–416

Russ GR, Alcala AC (2003) Marine reserves: rates and patterns of recovery and decline of predatory fish, 1983–2000. Ecol Appl 13:1553–1565

Russ GR, Alcala AC (2004) Marine reserves: long-term protection is required for full recovery of predatory fish populations. Oecologia 138(4):622–627

Russ GR, Alcala AC (2011) Enhanced biodiversity beyond marine reserve boundaries: the cup spillith over. Ecol Appl 21:241–250

Russ GR, Alcala AC, Maypa AP (2003) Spillover from marine reserves: the case of Naso vlamingii at Apo Island, the Philippines. Marine Ecol Prog Ser 264:15–20

Russ GR, Alcala AC, Maypa AP, Calumpong HP, White AT (2004) Marine reserve benefits local fisheries. Ecol Appl 14(2):597–606

Ruttenberg BI, Bohnsack JA, Ault JS, Smith SG, McClellan D, Javech J, Serafy JE (2013) Thirty years of change in reef fish communities in the Florida Keys: the importance of habitat, protection, and the long-view. Fish Res 144:115–116

Sabatier PA (1988) An advocacy coalition framework of policy change and the role of policy-orientated learning therein. Policy Sci 21(2–3):129–168

Sabatier PA (1998) The advocacy coalition framework: revisions and relevance for Europe. J Eur Public Policy 5(1):98–130

Sala E (2010) Talk given to a TED [Technology, Entertainment, Design] conference, part of Sylvia Earle's Mission Blue campaign. http://mission-blu.org.about/. Accessed 16 Dec 2010

Sale PF (2002) The science we need to develop for more effective management. In: Sale PF (ed) Coral reef fishes: dynamics and diversity in a complex ecosystem. Academic Press, San Diego

Sale PF et al [10 co-authors] (2005) Critical science gaps impede use of no-take fishery reserves. Trends Ecol Evol 20(2):74–80

Salm RV, Clark J, Siirila E (2000) Marine and coastal protected areas: a guide for planners and managers, 3rd edn. IUCN, Washington DC

Salomon AK et al [10 co-authors] (2011) Bridging the divide between fisheries and marine conservation science. Bull Marine Sci 87(2):251–274

Sarewitz D (2004) How science makes environmental controversies worse. Environ Sci Policy 7(5):385–403

Sarewitz D (2011, Oct 7) The voice of science: let's agree to disagree. Nature 478(7). doi:10.1038/478007a

Schlager E (1995) Policy-making and collective action- defining coalitions within the advocacy coalition framework. Policy Sci 28(3):243–270

Schopka SA (2007) Area closures in Icelandic waters and the real-time closure system. A historical review. Hafrannsoknastofnun Fjolrit 133:1–86

Sciberras M, Jenkins SR, Kaiser MJ, Hawkins SJ, Pullin AS (2013) Evaluating the biological effectiveness of fully and partially protected marine areas. Environ Evid 2(4):1–65

Scott J, Rachlow JL, Lackey RT, Pidgorna AB, Aycrigg JL, Feldman GR, Svancara LK, Rupp DA, Stanish DI, Steinhorst RK (2007) Policy advocacy in science: prevalence, perspectives, and implications for conservation biologists. Conserv Biol 21(1):29–35

Seafish (2009) The Seafish guide to marine protected reas. http://sin.seafish.org. Accessed 10 Dec 2009

Shears NT, Grace RV, Usmar NR, Kerr V, Babcock RC (2006) Long-term trends in lobster populations in a partially protected vs. no-take marine park. Biol Conserv 132(2):222–231

Shipp RL (2003) A perspective on marine reserves as a fishery management tool. Fisheries 28(12):10–21

Sloan NA (2002) History and application of the wilderness concept in marine conservation. Conserv Biol 16(2):294–305

Slooten E, Rayment W, Dawson S (2006) Offshore distribution of Hector's dolphins at Banks Peninsula, New Zealand: is the Banks Peninsula marine mammal sanctuary large enough? N. Z. J Marine Freshw Res 40(2):333–343

Smith RJ, Eastwood PD, Ota Y, Rogers SI (2009) Developing best practice for using Marxan to locate marine protected areas in European waters. ICES J Marine Sci 66(1):188–194

Smith M, Roheim C, Crowder L (2010) Sustainability and global seafood. Science 327:784–786

Spalding MD et al [14 co-authors] (2007) Marine ecoregions of the world: a bioregionalization of coastal and shelf areas. Bioscience 57(7):573–583

Spalding M, Meliane I, Milam A, Fitzgerald C, Hale LZ (2013) Protecting marine spaces: global targets and changing approaches. Ocean Yearb 27:213–248

Steel B, List P, Lach D, Shindler B (2004) The role of scientists in the environmental policy process: a case study from the American west. Environ Sci Policy 7(1):1–13

Steele J, Hoagland P (2004) Reply to Zeller and Russ. Fish Res 67(2):247–248

Stefansson G, Rosenberg AA (2006) Designing marine protected areas for migrating fish stocks. J Fish Biol 69(Supplement C):66–78

Stelzenmuller V, Pinnegar JK (2011) Monitoring fisheries effects of marine protected areas: current approaches and the need for integrated assessments. In: Claudet J (ed) Marine protected areas: a multidisciplinary approach. Cambridge University Press, Cambridge

Stelzenmuller V, Rogers S, Mills C (2008) Spatio-temporal patterns of fishing pressure on UK marine landscapes, and their implications for spatial planning and management. ICES J Marine Sci 65:1081–1091

Stelzenmuller V, Maynou F, Martin P (2009) Patterns of species and functional diversity around a coastal marine reserve: a fisheries perspective. Aquat Conserv-Marine Freshw Ecosyst 19(5):554–565

Stern P (2005) Deliberative methods for understanding environmental systems. Bioscience 55(11):976–982

Stewart RR, Noyce T, Possingham HP (2003) Opportunity cost of ad hoc marine reserve design decisions: an example from South Australia. Marine Ecol Prog Ser 253:25–38

Stewart GB, Côté IM, Kaiser MJ, Halpern BS, Lester SE, Bayliss HR, Mengersen K, Pullin AS (2008) Are marine protected areas effective tools for sustainable fisheries management? CEE Review 06-002 (SR 23). www.environmentalevidence.org/SR23.html. Accessed 19 June 2011

Stewart GB, Kaiser MJ, Côté IM, Halpern BS, Lester SE, Bayliss HR, Pullin AS (2009) Temperate marine reserves: global ecological effects and guidelines for future networks. Conserv Lett 2(6):243–253

Stokesbury KDE, Harris BP, Marino MC, Nogueira JL (2007) Sea scallop mass mortality in a marine protected area. Marine Ecol Prog Ser 349:151–158

Stone D (2002) Introduction: global knowledge and advocacy networks. Glob Netw 2(1):1–11

Sumaila UR, Guenette S, Alder J, Chuenpagdee R (2000) Addressing ecosystem effects of fishing using marine protected areas. ICES J Marine Sci 57:752–760

Sumpton WD, Jackson S (2010) Reproductive biology of snapper (*Pagrus auratus*) in subtropical areas of its range and management implications of reproductive differences with temperate populations. Asian Fish Sci 23(2):194–207

Sundström M (2000) A brief introduction: what is an epistemic community? http://www.svet.lu.se/joluschema/epistcomm.pdf. Accessed 13 Dec 2010

Suuronen P, Jounela P, Tschernij V (2010) Fishermen responses on marine protected areas in the Baltic cod fishery. Marine Policy 34:237–243

Sweeting C, Polunin NVC (2005) Marine protected areas for management of temperate North Atlantic fisheries: lessons learned in MPA use for sustainable fisheries exploitation and stock recovery. Report to Defra, London

Sweeting CJ, Badalamenti F, D'Anna G, Pipitone C, Polunin NVC (2009) Steeper biomass spectra of demersal fish communities after trawler exclusion in Sicily. ICES J Marine Sci 66(1): 195–202

Sweeting CJ, Armstrong I, Berney P, Armstrong N, Hardy MH (2011) Science-fishing industry partnership for assessing marine biodiversity: final report. Newcastle University. Report to the Marine Management Organisation

Symes D (2005) Fishing, the environment and the media. Fish Res 73(1–2):13–19

Symes D. Phillipson J (2009) Whatever became of social objectives in fisheries policy? Fish Res 95:1–5

Terauds A, Gales R, Baker GB, Alderman R (2006) Foraging areas of black-browed and grey-headed albatrosses breeding on Macquarie Island in relation to marine protected areas. Aquat Conserv-Marine Freshw Ecosyst 16(2):133–146

Thurstan R, Roberts C (2010) Ecological meltdown in the Firth of Clyde, Scotland: two centuries of change in a coastal marine ecosystem. PLoS ONE 5(7):1–14

Thurstan RH, Brockington S, Roberts CM (2010) The effects of 118 years of industrial fishing on UK bottom trawl fisheries. Nat Commun 1:15

Tissot B, Walsh W, Stevenson T (2013) The effects of a network of MPAs in Hawaii to promote sustainability in an aquarium fishery. Fish Res 144:115

Tolimieri N, Andrews KS, Williams GD, Katz SL, Levin PS (2009) Home range size and patterns of space use by lingcod, copper rockfish and quillback rockfish in relation to diet and tidal cycles. Marine Ecol Prog Ser 380:229–243

Tomkins J, Kotiaho J (2004) Publication bias in meta-analysis: seeing the wood for the trees. Oikos 104(1):194–196

Toropova C, Kenchington R, Vierros M, Meliane I (2010) Benefits and challenges of MPA strategies. In: Toropova C, Meliane I, Laffoley D, Matthews E Spalding M (eds) Global ocean protection: present status and future possibilities. Agence des Aires Marines Protégées. Brest, France

Turnhout E, Hisschemoller M, Eijsackers H (2007) Ecological indicators: between the two fires of science and policy. Ecol Indic 7(2):215–228

Underwood AJ (1993) The mechanics of spatially replicated sampling programmes to detect environmental impacts in a variable world. Aust J Ecol 18(1):99–116

UNEP (2011) Taking steps towards local and coastal ecosystem-based management- an introductory guide. UNEP regional seas reports and studies No. 189

Vandeperre F et al [24 co-authors] (2011) Effects of no-take area size and age of marine protected areas on fisheries yields: a meta-analytical approach. Fish Fish 12:412–426

Vanderklift MA, Ward TJ, Phillips JC (1998) Use of assemblages derived from different taxonomic levels to select areas for conserving marine biodiversity. Biol Conserv 86(3):307–315

Vincent MA, Atkins SM, Lumb CM, Golding N, Lieberknecht LM, Webster M (2004) Marine nature conservation and sustainable development—the Irish Sea Pilot. Report to Defra by the JNNC, Peterborough

Voyer M, Gladstone W, Goodall H (2012) Methods of social assessment in marine protected area planning: is public participation enough? Marine Policy 36(2):432–439

Wagner LD, Ross JV, Possingham HP (2007) Catastrophe management and inter-reserve distance for marine reserve networks. Ecol Model 201:82–88

Wantiez L, Tholliot P, Kulbicki M (1997) Effects of marine reserves on coral reef fish communities from five islands in New Caledonia. Coral Reefs 16(4):215–224

Watling L, Norse EA (1998) Disturbance of the seabed by mobile fishing gear: a comparison to forest clearcutting. Conserv Biol 12(6):1180–1197

Weible CM (2007) An advocacy coalition framework approach to stakeholder analysis: understanding the political context of California marine protected area policy. J Public Adm Res Theory 17(1):95–117

Weible CM (2008) Caught in a maelstrom: implementing California marine protected areas. Coast Manage 36(4):350–373

Weible CM, Sabatier PA (2005) Comparing policy networks: marine protected areas in California. Policy Stud J 33(2):181–201

Weible C, Sabatier PA, Lubell M (2004) A comparison of a collaborative and top-down approach to the use of science in policy: establishing marine protected areas in California. Policy Stud J 32(2):187–207

Wells S (2014) Perspective: designating marine conservation zones in England—a phased approach. MPA News 15(4):7–8 (March/April 2014)

Wenzel L, Brock R (2013) Key findings from Fisheries Research: marine protected areas as a fisheries management tool. Office of National Marine Sanctuaries. NOAA, Washington DC

Wenzel L, Gass J, D'Iorio M, Blackburn J (2013) A national perspective on the role of marine protected areas in sustaining fisheries. Fish Res 144:23–27

White C, Costello C (2014) Close the high seas to fishing. PLOS Biol 12(3):1–5

Wiersma Y, Nudds T (2012) Percentage conservation targets are problematic for marine mammals. Proc Natl Acad Sci USA 109(6):E288

Willis TJ, Parsons DM, Babcock RC (2001) Evidence for long-term site fidelity of snapper (*Pagrus auratus*) within a marine reserve. N. Z. J Marine Freshw Res 35(3):581–590

Willis TJ, Millar RB, Babcock RC, Tolimieri N (2003a) Burdens of evidence and the benefits of marine reserves: putting Descartes before des horse? Environ Conserv 30(2):97–103

Willis TJ, Millar RB, Babcock RC (2003b) Protection of exploited fish in temperate regions: high density and biomass of snapper *Pagrus auratus* (Sparidae) in northern New Zealand marine reserves. J Appl Ecol 40(2):214–227

Wood LJ, Fish L, Laughren J, Pauly D (2008) Assessing progress towards global marine protection targets: shortfalls in information and action. Oryx 42(3):340–351

Worm B et al [13 co-authors] (2006) Impacts of biodiversity loss on ocean ecosystem services. Science 314(5800):787–790

Worm B et al [20 co-authors] (2009) Rebuilding global fisheries. Science 325:578–585

Wright D (2010) The ocean planet: a proposal for fundamental changes in marine management. Marinet. www.marinet.org.uk. Accessed 12 Dec 2010

WWF (2010) High seas MPAs: regional approaches and experiences. Background document for the side event at the 12th UNEP global meeting of the Regional Seas Conventions and Action Plans, Bergen, 20–22 Sept 2010

Zeller D (1997) Home range and activity patterns of the coral trout Plectropomus leopardus (Serranidae). Marine Ecol Prog Ser 154:65–77

Index

A. Caveen et al., *The Controversy over Marine Protected Areas*,
SpringerBriefs in Environmental Science, DOI 10.1007/978-3-319-10957-2

MIX
Papier aus verantwortungsvollen Quellen
Paper from responsible sources
FSC® C105338
FSC
www.fsc.org

If you have any concerns about our products,
you can contact us on
ProductSafety@springernature.com

In case Publisher is established outside the EU,
the EU authorized representative is:
Springer Nature Customer Service Center GmbH
Europaplatz 3, 69115 Heidelberg, Germany

Printed by Libri Plureos GmbH
in Hamburg, Germany